Modeling and partially coordinated diagnosis of asynchronous discrete-event systems

Dissertation zur Erlangung des Grades eines

Doktor-Ingenieurs

der Fakultät für Elektrotechnik und Informationstechnik
an der Ruhr-Universität Bochum

Vorgelegt von

Sebastian Drüppel
geboren in Dortmund

Bochum, Juli 2011

1st Reviewer: Prof. Dr.-Ing. Jan Lunze,
Ruhr-Universität Bochum

2nd Reviewer: Univ.-Prof. Dr.-Ing. Alexander Fay
Helmut-Schmidt-Universität Hamburg

Thesis submitted on: 05.07.2011
Date of examination: 03.11.2011

Bibliografische Information der Deutschen Nationalbibliothek

Die Deutsche Nationalbibliothek verzeichnet diese Publikation in der Deutschen Nationalbibliografie; detaillierte bibliografische Daten sind im Internet über http://dnb.d-nb.de abrufbar.

ISBN 978-3-8325-3259-8

Logos Verlag Berlin GmbH
Comeniushof, Gubener Str. 47,
10243 Berlin
Tel.: +49 (0)30 42 85 10 90
Fax: +49 (0)30 42 85 10 92
INTERNET: http://www.logos-verlag.de

Acknowledgements

This thesis is the result of three years of research at the *Corporate Sector Research and Advanced Engineering* of the Robert Bosch GmbH in cooperation with the *Institute of Automation and Computer Control*, Prof. Jan Lunze, at the Ruhr-Universität Bochum and a sabbatical leave I took from my work at the RWE Power AG to write down the results.

I would like to express my deepest gratitude to Prof. Jan Lunze for his excellent guidance, his support and encouragement. I must especially thank him for having me at his institute during my sabbatical leave. For reviewing this work, I wish to cordially thank Prof. Alexander Fay.

Special thanks go to Dr. Martin Fritz who cleared the way for this work both at the Robert Bosch GmbH and the Ruhr-Universität Bochum. I am indebted to Steffen Ballmann who gave me the idea of taking a sabbatical leave to finish my thesis and to Ferdinand Scheiff whose advocacy made it possible to do so at the RWE Power AG.

A great thank you goes to my former colleagues at the Ruhr-Universität Bochum and at the Robert Bosch GmbH. I would especially like to thank Dr. Florian Wenck, Dr. Jörg Neidig and Yannick Nke for the fruitful discussions on discrete event systems and the latter for reviewing this thesis. I wish to thank Kosmas Petridis for the nice coffee breaks with discussions on anything but discrete event systems.

Several students have worked with me to tackle the various problems solved in this thesis. I appreciate having had those capable minds helping me. Especially the engagement of Yannick Nke and Christian Maul will stay in my mind as we have spend many hours on discussing tasks associated with this work.

Personally, I wish to express me heartfelt gratitude to my parents, my sister, my parents-in-law, my grandparents and my friends for their affection and support. At last but certainly not the least I would like to express my deepest appreciation to my wife Nadine, who accompanied me with her love and kindness.

Thank you!

Hannover, October 2012 Sebastian Drüppel

Contents

Abstract

This thesis presents a novel approach to modeling, analysis and diagnosis of coupled mechatronical systems with *partially autonomous* behavior and *asynchronous* state transitions. The systems under consideration have the following properties: The internal interactions are immeasurable but reliable and the measurements relevant for diagnosis are given as a sequence of events even though the underlying control is assumed to be continuous.

Asynchronous networks of input/output automata (I/O-automata) are developed in this thesis to cope with partial coupling between components and to reduce the *computational complexity* of the diagnostic algorithms. I/O-automata are used to model those components. Faults are modeled as an additional parameter of their behavioral relation. The components' measurable inputs and outputs are modeled as control signals. Interconnection signals are used to model the internal dependencies among the components. They are linked via an interaction block to one another. The empty symbol ε is introduced to model the non-interaction between components that operate independently such that the sequences of all local measurements are of equal length. Hence, the behavior of all components is synchronized upon the occurrence of events. This modeling concept is based on the parallel composition known from standard automata. It is shown that it is applicable for at least the same class of asynchronous discrete-event systems as the known modeling formalism while having certain advantages.

To ensure the *well-posedness* of the developed models, the criterion known from synchronous networks of I/O-automata is extended, since the coupling of component models can result in loops. To check for partially autonomous behavior, two types of autonomy are introduced. Conditions for *structural autonomy* are derived based on a graph theoretical analysis of the corresponding directed graph. *State-dependent autonomy* is investigated in terms of the behavioral relation and used for simplifications in diagnosis. Two algorithms for the *simulation* of the behavior of asynchronous networks of I/O-automata with identical results are presented. The first one uses the whole equivalent I/O-automaton which has been pre-computed by composition, whereas the second one reduces the model size by applying the composition rule only to the part of the behavioral relation concerning the actual state and inputs. This approach will be referred to as online composition.

To carry out the diagnosis, three different information structures are investigated: Centralized, decentralized and partially coordinated. The centralized approach yields the ideal diagnostic result, but reduction of the computational complexity by using online composition is rather small. It has already been applied to synchronous networks of I/O-automata. The modeling introduced in this thesis makes it applicable to both, asynchronous and synchronous networks of I/O-automata as composition leads to the loss of network effects. To reduce the computational complexity further, the diagnostic task is solved in a decentralized way next. The locally obtained diagnostic results are synchronized upon the occurrence of events. They are used to get the decentrally obtained diagnostic result which is *complete* and *sound* for well-posed networks in the case of state-dependent autonomy. A lack of soundness arises in the general case whereas the completeness property is still given. To overcome the soundness issue, *partially coordinated diagnosis* is introduced: The ideal diagnostic result is obtained by removing contradictions from the locally estimated sets of interconnection signals and faults in a centralized coordinator. The result of coordination has not to be sent back to the local diagnostic units for well-posed networks in the case of state-dependent autonomy as opposed to the bidirectional algorithm known from synchronous networks of I/O-automata. The partially coordinated diagnostic result has to be stored in the coordinator for the use in the next time step. The developed diagnostic algorithm is applicable to synchronous networks of I/O-automata due to the use of the empty symbol.

Deutsche Zusammenfassung

Moderne technische Systeme sind hoch komplex und weithin automatisiert. Die hohe Komplexität steht in direktem Zusammenhang mit der Anzahl interagierender Komponenten, die eine Vielzahl an Informationen austauschen. Unvorhersehbare *Fehler* können das Verhalten eines Systems schwerwiegend ändern, so dass die angestrebte Funktionsweise gar nicht oder nicht mit der erforderlichen Genauigkeit erreicht werden kann. Um eine zuverlässige Funktionsweise des Systems zu gewährleisten und kritischen Situationen vorzubeugen, müssen Fehlfunktionen schnellst möglich erkannt und isoliert werden. Die wahren Gründe für fehlerhaftes Verhalten können im Allgemeinen von menschlichen Anlagenführern nicht identifiziert werden, weil Fehler typischerweise zu nicht zuordenbaren Auswirkungen führen. Deshalb besteht die Notwendigkeit der Entwicklung von Diagnosemethoden, die Fehler in komplexen Systemen zuverlässig detektieren und identifizieren.

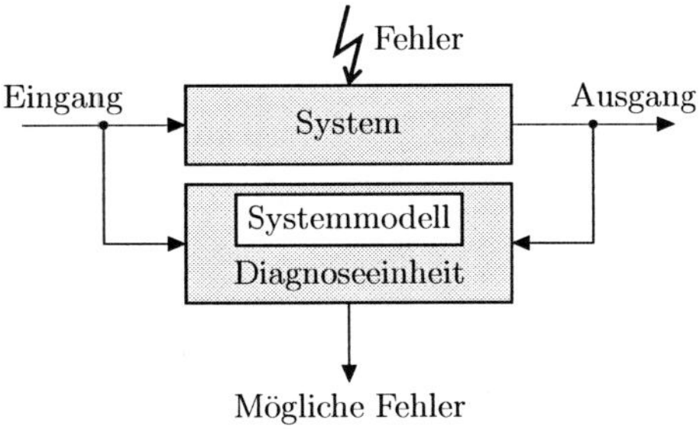

Abbildung 1: Grundkonzept der Diagnose

Diese Arbeit behandelt die *modellbasierte* Diagnose eines technischen Systems (Abbildung 1). Die Interaktion des betrachteten Systems mit seinem Umfeld wird durch Eingangs- und Ausgangssignale beschrieben. Es wird angenommen, dass der *unbekannte* Fehler das Eingangs-/ Ausgangsverhalten (E/A-Verhalten) des Systems grundlegend verändert. Die Diagnoseaufgabe wird gelöst, indem Messungen der Eingangs- und der Ausgangssignale mit einem mathematischen Modell des Systems verglichen werden.

Das Diagnosesystem ist modular aufgebaut: Die Systembeschreibung und der Diagnosealgorithmus sind klar voneinander getrennt. Änderungen am System erfordern lediglich Anpassungen des Modells. Der Diagnosealgorithmus als solcher bleibt davon unberührt.

Die folgenden Fragen werden in dieser Arbeit beantwortet:

1. Modellbildung: Wie können komplexe technische Systeme zu Diagnosezwecken modelliert werden?

2. Diagnose: Wie sieht eine zur Systembeschreibung passende Diagnosemethode aus?

Für die Diagnose *gekoppelter mechatronischer Systeme* als Spezialfalls komplexer technischer Systeme werden in dieser Arbeit verschiedene Verfahren entwickelt. Die betrachtete Systemklasse hat die folgenden Eigenschaften: Ein System besteht aus einer endlichen Menge interagierender Komponenten mit einer nichtlinearen bzw. hybriden Dynamik. Die Interaktion zwischen den Komponenten ist im Allgemeinen nicht messbar, aber verlässlich. Die für die Diagnose relevanten Messungen werden als diskrete Wertefolge angegeben, obwohl angenommen wird, dass unterlagerte Regelungen kontinuierlich arbeiten. In verschiedenen Arbeitspunkten können einige aber nicht notwendigerweise alle Komponenten interagieren, um eine gewünschte Aufgabe zu erfüllen. Diese Komponenten werden *partiell autonom* arbeitend genannt, d.h. sie ändern ihren Zustand während die übrigen Komponenten in ihrem aktuellen Zustand verharren. Somit bewegen sich diese Komponenten *asynchron* im Vergleich zum restlichen System.

In dieser Arbeit sind **Methoden zur Komplexitätsreduktion** der ereignisdiskreten Modellierung und der Algorithmen zur Diagnose gekoppelter mechatronischer Systeme mit partiell autonomem Verhalten entwickelt worden. Im Einzelnen sind zu den folgenden Punkten Beiträge erarbeitet worden:

1. Komponentenorientierte Modellierung. Im Bereich der Modellierung asynchroner, ereignisdiskreter Systeme sind die folgenden Ergebnisse erzielt worden:

- *Asynchrone Eingangs-/Ausgangsautomatennetze* sind entwickelt worden, um teilweise Autonomie in der Arbeitsweise mechatronischer Systeme zu beschreiben. Dieser Ansatz, deren Vorstudie sich in [2] befindet, ist in [5] veröffentlicht worden und wird im Kapitel 3 vorgestellt. Eingangs-/Ausgangsautomaten (E/A-Automaten) werden, wie in der Abbildung 2 dargestellt, als Komponentenmodelle $\mathcal{N}_i$ verwendet. Die messbaren Ein- und Ausgänge der Komponenten werden als Steuersignale v_i und w_i modelliert. Die Interaktion zwischen den Komponenten wird über die in dieser Arbeit eingeführten Koppelsignale s_i und r_i beschrieben. Der Koppelblock $\boldsymbol{K}$ verbindet die Koppelsignale der Komponenten miteinander. Somit wird die Ursache-Wirkungsbeziehung des Systems explizit im Modell abgebildet. Änderungen im System können einfach in das Modell übernommen werden. Ein zum E/A-Automatennetz äquivalenter E/A-Automat kann mit Hilfe des im Abschnitt 3.5.2 angegebenen Kompositionsverfahrens gebildet werden.

- Fehler werden auf Grund der Annahmen, dass sie das Systemverhalten gravierend ändern, als zusätzlicher Parameter der Verhaltensrelation eines E/A-Automaten modelliert [20]. Die Zusammenführung all jener Modelle, die jeweils das Verhalten des

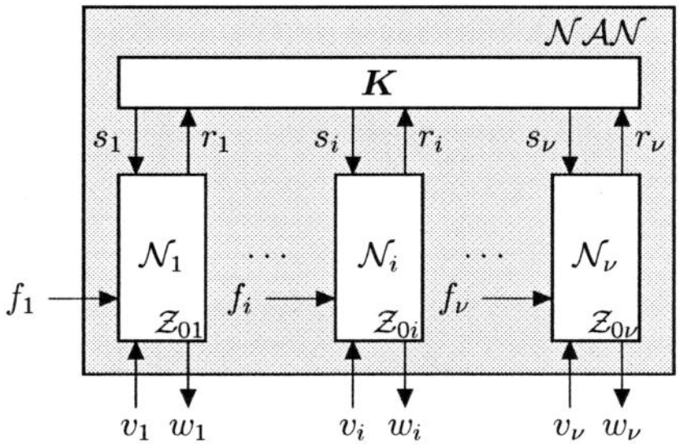

Abbildung 2: Asynchrones Eingangs-/Ausgangsautomatennetz

Systems unter der Annahme eines Fehlers beschreiben, liefert das gesamte Systemmodell. Somit entsteht ein Multi-Modell zur Beschreibung des Systemverhaltens.

- Der neue Modellierungsansatz wird im Abschnitt 3.6 zur Beschreibung asynchroner Zustandsübergänge durch die Einführung des leeren Symbols ε erweitert [5, 72]. Dieses Symbol wird verwendet, um die Nicht-Interaktion zwischen Komponenten zu beschreiben, die zur Erfüllung einer gewünschten Aufgabe eigenständig arbeiten. Durch diese Modellerweiterung wird sichergestellt, dass alle lokalen Messungen die gleiche Anzahl an Elementen aufweisen. Somit ist das Verhalten aller Komponenten auf das Auftreten von Ereignissen im System synchronisiert.
- Das neue Modellierungskonzept basiert auf der von Standardautomaten bekannten parallelen Komposition [27, 70], die die Kopplungen zwischen den Komponenten implizit durch die Bezeichnung der Ereignisse beschreibt. Im Abschnitt 3.8 und im Beitrag [5] wird gezeigt, dass die neue Modellform mindestens auf die gleiche Klasse asynchroner, ereignisdiskreter Systeme anwendbar ist wie gekoppelte Standardautomaten, jedoch dem gegenüber Vorteile bietet. Der formale Beweis der Äquivalenz befindet sich in [73].

2. Analyse. Die Analyse asynchroner Eingangs-/ Ausgangsautomatennetze umfasst die folgenden Punkte:

- Die Verschaltung von Komponenten kann im Allgemeinen zu algebraischen Schleifen führen, d.h. das Verhalten einer Komponente direkt und unmittelbar über einen Signalpfad im Netzwerk von sich selbst abhängt. Diese Schleifen können teilweise nicht eindeutig lösbar sein, so dass das Netz nicht *wohl-definiert* ist. Dieses Problem ist von synchronen E/A-Automatennetzen aus [84, 86] bekannt. In der Diplomarbeit [11] und der daraus resultierenden Veröffentlichung [6] ist deren Kriterium auf einer algorithmischen Basis für die Analyse deterministischer, asynchroner E/A-Automatennetze erweitert worden. Die dazugehörige mathematische Beschreibung wird im Abschnitt 4.1 eingeführt und auf den nichtdeterministischen Fall erweitert. Es wird gezeigt, dass der durch Komposition eines deterministischen, asynchronen E/A-Automatennetzes erhaltene äquivalente E/A-Automat entweder deterministisch (Theorem 1) oder schwach-deterministisch (Theorem 2) sein kann, sowie nichtdeterministisch für nichtdeterministische, asynchrone E/A-Automatennetze (Theorem 3).

- Zwei Arten von Autonomie werden in dieser Arbeit eingeführt. Kriterien für *strukturelle Autonomie* werden im Abschnitt 4.2.1 basierend auf einer Suche nach nicht miteinander verbundenen Mengen stark zusammenhängender Knoten im korrespondierenden gerichteten Grafen entwickelt. Für den Fortgang dieser Arbeit wird angenommen, dass eine Unterteilung des betrachteten Systems in nicht miteinander verbundene Mengen stark zusammenhängender Knoten nicht möglich ist, so dass das System im Hinblick auf die Diagnose als Ganzes analysiert werden muss. *Zustandsabhängige Autonomie* wird im Abschnitt 4.2.2 anhand der Verhaltensrelation untersucht. Sie bildet die Grundlage für Vereinfachungen in der Diagnose des betrachteten Systems und ist in [72] für eine reduzierte Modellform ohne algebraische Schleifen veröffentlicht.

- Zwei Algorithmen zur *Simulation* des Verhaltens asynchroner E/A-Automatennetze werden im Abschnitt 4.3 unter Verwendung der in der Abbildung 3 dargestellten Struktur präsentiert. Beide Algorithmen liefern das gleiche Ergebnis (Theorem 4). Der erste Algorithmus verwendet den vollständigen äquivalenten E/A-Automaten, der im Vorfeld mittels Komposition berechnet worden ist, wohingegen der Zweite die Modellgröße reduziert, in dem online die Kompositionsregel auf den Teil der Verhaltensrelation angewendet wird, der den aktuellen Zustand und Eingang betrifft. Dieser Ansatz wird als *online Komposition* bezeichnet.

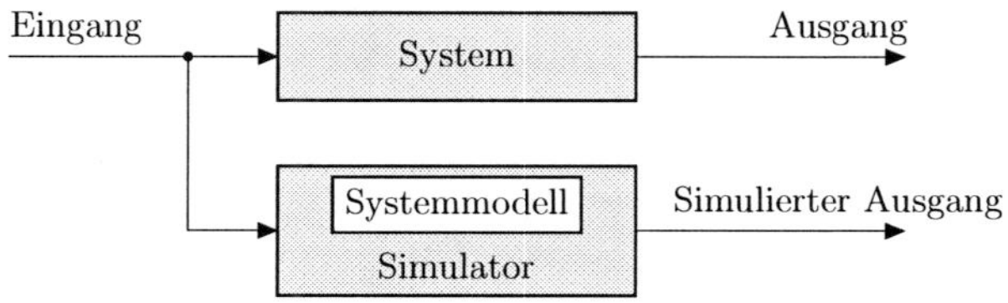

Abbildung 3: Simulation gekoppelter, ereignisdiskreter Systeme

3. Partiell koordinierte Diagnose. Die folgenden Beiträge zur Diagnose asynchroner, ereignisdiskreter Systeme bilden die Hauptergebnisse dieser Arbeit:

- Im Rahmen der Entwicklung der in dieser Arbeit präsentierten Diagnoseverfahren konnte die Gleichheit der zentralen und dezentralen Diagnose bewiesen werden, wenn Fehler als zusätzlicher Parameter der Verhaltensrelation von Standardautomaten, die mittels paralleler Komposition gekoppelt sind, modelliert werden. Dieser Ansatz ist in

[1, 3] veröffentlicht und in [4] zur modularen Diagnose eines Fahrdynamikregelungssystems angewendet worden. Da er nicht auf den hier betrachteten E/A-Automaten basiert, wird er als Zwischenergebnisses angesehen und hier nicht weiter behandelt.

- Als Grundlage der hier behandelten Diagnoseverfahren wird im Abschnitt 5.1 eine Methode zur *Zustandsbeobachtung* analysiert. Da die Zustandsbeobachtung zum Verständnis der Diagnose notwendig ist, jedoch nicht den Fokus dieser Arbeit bildet, wird sie für eine zentrale Informationsstruktur betrachtet. Die Gleichheit des zentral ermittelten Beobachtungsergebnis mit dem idealen Beobachtungsergebnis wird gezeigt: $\mathcal{Z}^c(k) = \mathcal{Z}^\star(k)$ (Theorem 5).

- Die Methode zur Zustandsbeobachtung wird im Abschnitt 5.2 für die Diagnose eines E/A-Automaten unter Verwendung einer zentralen Informationsstruktur erweitert. Der Diagnosealgorithmus liefert das ideale Diagnoseergebnis: $\mathcal{F}^c(k) = \mathcal{F}^\star(k)$, löst jedoch nicht das Komplexitätsproblem. Die online Komposition der Verhaltensrelation im Zusammenhang mit einer zentralen Diagnoseeinheit aus [78] wird im Abschnitt 5.3 auf asynchrone E/A-Automatennetze erweitert. Dieser Ansatz liefert ebenfalls das ideale Diagnoseergebnis (Theorem 6), löst das Komplexitätsproblem jedoch nur bedingt. Er wird im Rahmen dieser Arbeit als Referenz verwendet.

- Zur weiteren Komplexitätsreduktion wird die Dezentralisierung der Diagnoseaufgabe aus [85] auf das neu entwickelte Modellierungskonzept angewendet (Abbildung 4).

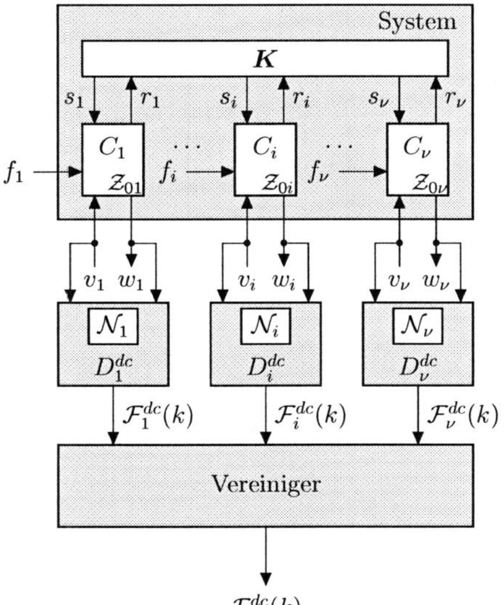

Abbildung 4: Dezentrale Diagnose gekoppelter, ereignisdiskreter Systeme

Diese dezentrale Diagnosemethode ist online anwendbar und hat die geringste Komplexität hinsichtlich benötigter Speicherkapazität und Rechenzeit. Sie wird im Kapitel 6 für den allgemeinen Fall behandelt, eine Vereinfachung, die Abhängigkeiten zwischen den Komponenten eines Systems ausschließt, ist in [72] veröffentlicht worden. In lokalen Diagnoseeinheiten D_i^{dc} wird eine Erweiterung des zentralen Diagnosealgorithmus verwendet, um lokal das ideale Diagnoseergebnis zu berechnen: $\mathcal{F}_i^{dc}(k) = \mathcal{F}_i^{\star}(k)$ (Theorem 8). Daraus wird im „Vereiniger" das globale Diagnoseergebnis $\mathcal{F}^{dc}(k)$ berechnet. Dieses ist in wohl-definierten Netzen gleich dem idealen Diagnoseergebnis, wenn Zustands-abhängige Autonomie vorliegt (Theorem 11). Generell geht die Richtigkeit des Diagnoseergebnisses verloren (Theorem 10), weil die Abhängigkeiten zwischen den Komponenten bei der dezentralen Diagnose vernachlässigt werden. Das dezentral ermittelte Diagnoseergebnis verschlechtert sich also im Allgemeinen: $\mathcal{F}^{dc}(k) \supseteq \mathcal{F}^{\star}(k)$ (Theorem 9).

- Um die Richtigkeit des Diagnoseergebnisses unter Zuhilfenahme eines strukturierten Modells zu erlangen, wird eine Methode zur *partiell koordinierten Diagnose* (Abbildung 5) in Kapitel 7 eingeführt.

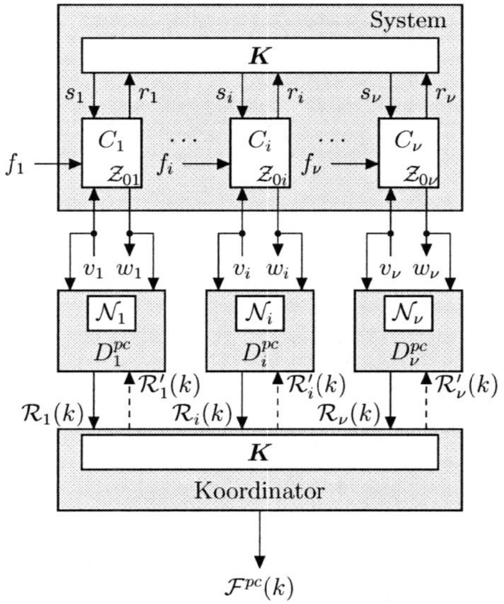

Abbildung 5: Partiell koordinierte Diagnose gekoppelter, ereignisdiskreter Systeme

Dazu wird das bidirektionale Koordinationskonzept aus [87, 88] auf asynchrone E/A-Automatennetze erweitert: In den lokalen Diagnoseeinheiten D_i^{pc} werden die Mengen $\mathcal{R}_i(k)$ der möglichen Paare von Koppeleingang $s_i(k)$, Koppelausgang $r_i(k)$ und Fehler f_i in jedem Zeitschritt k ermittelt und an den zentralen Koordinator gesendet.

Der Koordinator entfernt aus diesen Mengen Widersprüche zur Berechnung des partiell koordinierten Diagnoseergebnisses $\mathcal{F}^{pc}(k)$. Dieses ist in jedem Zeitschritt gleich dem idealen Diagnoseergebnis: $\mathcal{F}^{pc}(k) = \mathcal{F}^{\star}(k)$ (Theorem 12). In dieser Arbeit wird gezeigt, dass der Koordinator entgegen den Aussagen in [84] einen Datenspeicher braucht, um das partiell koordinierte Diagnoseergebnis $\mathcal{F}^{pc}(k)$ für die Verwendung im nächsten Zeitschritt zu speichern.
Das Vorliegen von Zustands-abhängiger Autonomie führt zu den folgenden Vereinfachungen: Die lokalen Diagnoseeinheiten der Komponenten, die Zustands-autonom arbeiten, erhalten im aktuellen Zeitschritt vom Koordinator keine Daten $\mathcal{R}_i'(k)$. Stellt der Koordinator in einem Zeitschritt fest, dass alle Komponenten Zustands-autonom arbeiten, ist eine Koordination in wohl-definierten Netzen nicht notwendig. In diesem Fall erhalten alle lokalen Diagnoseeinheiten keine Daten vom Koordinator.
Der hier vorgestellte partiell koordinierte Diagnosealgorithmus ist online anwendbar. Er reduziert die algorithmische Komplexität deutlich. Allerdings ist bedingt durch die Koordination die benötigte Rechenzeit höher als die des dezentralen Ansatzes. Durch die Verwendung des leeren Symbols kann er auf synchrone E/A-Automatennetze angewendet werden.

4. Anwendungsbeispiele. Zwei Anwendungsbeispiele werden im Kapitel 8 vorgestellt. Die Anwendung des dezentralen und des partiell koordinierten Diagnosealgorithmus auf eine verfahrenstechnische Anlage werden im Abschnitt 8.1 vorgestellt. Diese Ergebnisse basieren auf der Studienarbeit [9]. Die Simulationsmethode wird auf die Sortieranlage aus [70] im Abschnitt 8.2 angewendet. Die vorgestellten Ergebnisse sind in der Diplomarbeit [11] erzielt worden.

5. Implementierung. Eine MATLAB Toolbox ist zur Modellierung, Analyse und Diagnose asynchroner E/A-Automatennetze entwickelt worden. Der erste Ansatz ist im Rahmen des Praktikums [10] für gekoppelte Standardautomaten entstanden. Er ist in der Diplomarbeit [11] zur Modellierung der in den Abschnitten 3.2 bis 3.6 eingeführten, deterministischer E/A-Automatennetze und deren im Abschnitten 4.1 beschriebenen Analyse erweitert worden. Die Erweiterung zur Behandlung des nichtdeterministischen Falls ist im Rahmen des Praktikums [7] begonnen worden. Im Rahmen der Praktika [8] sind diese Erweiterung sowie die Erweiterung zur Behandlung von Fehlern als zusätzlicher Parameter der Verhaltensrelation fertig gestellt worden. Zusätzlich sind eine grafische Benutzeroberfläche sowie die im Kapitel 6 beschriebene Methode zur zentrale Diagnose implementiert worden. Die abschließenden Arbeiten sind in der Studienarbeit [9] zur Realisierung der Methode zur partiell koordinierten Diagnose durchgeführt worden.

Chapter 1

Introduction

1.1 Diagnosis of complex engineering systems

Modern technological systems are highly complex and widely automated. The high complexity arises from the large number of interacting components sharing a large amount of information. Unpredictable *faults* may change the behavior of a system severely such that the designed operational strategies cannot be fulfilled with the required performance or may even fail. To ensure dependability of the system operation and to avoid critical situations, such malfunctions need to be detected and isolated as soon as possible. Solely based on the input/output measurements, finding the true cause of an abnormal behavior is, in general, impossible for humans observing the system because faults mostly cause ambiguous symptoms. Hence, there is a need to develop diagnostic components which are capable of detecting and identifying faults reliably.

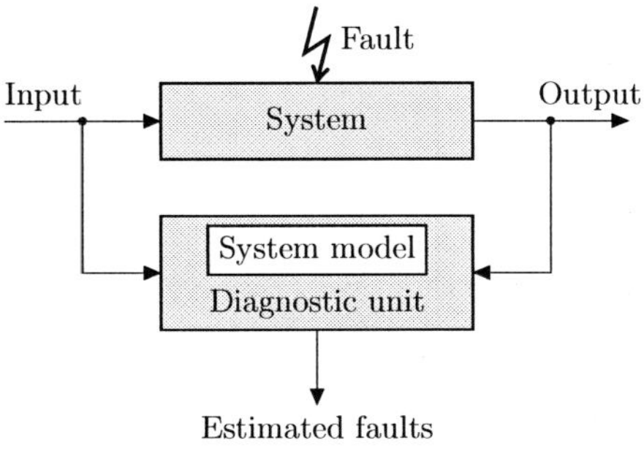

Figure 1.1: Basic concept of diagnosis

This thesis deals with *model-based* diagnosis (Fig. 1.1). The interaction of a system under consideration with its environment is described by input and output signals. It is assumed that the *unknown* fault signal alters the input/output-behavior (I/O-behavior) of this system. Given a mathematical description of the system, the diagnostic task is solved by comparing measurements of the input and the output with this model to find the fault.

The diagnostic system is designed in a modular way: The system description and the diagnostic algorithm are separated clearly from each other. Hence, changes in the system do only result in adjusting of the model. The diagnostic algorithm remains unchanged and the overall diagnostic system retains its structure. Answers to the following questions will be given in the sequel:

1. Modeling: How to model complex engineering systems for diagnostic purposes?

2. Diagnosis: How to design a diagnostic method based on this system description?

The diagnostic approaches are developed in this thesis for diagnosis of *coupled mechatronical systems* which are a special class of complex engineering systems. They consist of a set of interconnected components which may be nonlinear or hybrid in nature, i.e. with a discretely switching continuous dynamic. Their design strategy may require different operating points, where several but not necessarily all components are interacting to fulfill a desired task. Those components are said to work *partially autonomously* because they change their states whereas the remaining components stay in their current states. Thus, there are *asynchronous* state changes in the system. In general, internal interactions among the components are assumed to be immeasurable, but reliable. Besides this, most mechatronical systems are operated under changing environmental conditions, like temperature or moisture, which may alter the system in a way that its behavior deviates within a given tolerable range.

Example 1 *Vehicle dynamics management (VDM)*

Figure 1.2 depicts the Bosch approach for coordinating vehicle dynamics functions by integrated control of active chassis systems [12, 59, 103, 125, 126, 127]. The task of the VDM is to stabilize the vehicle in critical driving situations, such as over- and under-steering and to improve its agility. The VDM is considered in this work as a typical representative of the aforementioned class of systems because of the following:

1. *It consists of a set of components that are partially coupled.*
2. *It has a discrete-event character based on qualitative measurements.*
3. *It needs to be diagnosed reliably during all driving situations.*

1. Partially coupling. *The core of the VDM is the yaw rate controller which evaluates the actual driving situation. If a critical driving situation has been detected, a stabilizing target yaw torque is distributed to either the active front steering (AFS), the electronic stabilization programm (ESP), the active roll control (ARC) and / or the drive train management according to a certain coordination strategy. The components*

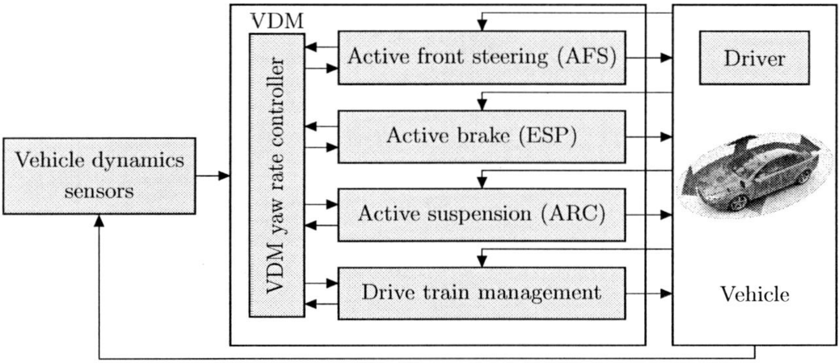

Figure 1.2: The vehicle dynamics management system VDM [59, 125, 126, 127]

calculate stabilizing improvements depending on their functionality from this target value. As vehicles are usually operated in uncritical situations, the components of the VDM work predominantly autonomous and can behave asynchronously, i.e. move without any influence of other components. The coupling between the components will only be activated in critical driving situations.

2. Discrete-event character of mechatronical systems. *The operational strategy bases on commands like "Build-up a stabilizing brake torque" sent from the yaw rate controller to the ESP or the corresponding answer "Stabilizing brake torque adjusted". Thus, the relevant measurements can be given as a sequence of events (Section 3.1) even though the underlying control is continuous. Input/output automata (I/O-automat) have been proven to be suitable for diagnosis because only relevant information of the system is considered for diagnosis. Systems consisting of a set of partially coupled components suggest to use the asynchronous networks of I/O-automata introduced in Chapter 3 because they reduce the* ***computational complexity****. This problem is known as state space explosion, i.e. the state space of an I/O-automaton grows exponentially with the number of components such that the resulting monolithic model becomes quickly unmanageable even with modern computers.*

3. Need for reliable diagnosis. *This item is obvious because malfunction may result e.g. in a crash of the vehicle which must be avoided. But the constraints on the diagnostic system to work always and in parallel to the observed system are hard [60, 89]. Online applicability, i.e. a low amount of storage capacity and computation time, is reached by* ***partially coordinated diagnosis*** *developed in Chapter 7. It relies on the principle of consistency-based diagnosis introduced in Chapter 2 and accounts for the system structure with the partial interaction between the components.*

1.2 Aim and tasks

The aim of this thesis is to develop an approach to detect and isolate faults reliably in coupled mechatronical systems. The solution presented here is based on the concept of *model-based diagnosis.* It can be split up according to Fig. 1.1 into three major tasks to answer the questions stated in Section 1.1.

1. Component-oriented modeling. A suitable model for this class of systems has to be set up. It has been pointed out in Section 1.1 that a qualitative description in terms of event sequences is well-suited for diagnostic purposes. The concept of nondeterministic I/O-automata will be used here because it explicitly covers the cause-and-effect-chains of the system. Standard automata and Petri nets suffer from the problem of implicitly describing the interrelations of the system and the fact that the state transitions are represented by events rather than input and output symbols. They will not be considered in the remainder.

The complexity of most mechatronical systems prohibits the use of a monolithic description by one automaton due to the problem of state space explosion. Hence, there is a need to develop a new modeling formalism that explicitly describes the cause-and-effect-chains of the system on a component-oriented basis and allows for asynchrony of several components. The solution based on *asynchronous networks of I/O-automata* will be presented in Chapter 3 in detail including the design of composition operators to obtain the behavior of the system based on the component models. It is a bottom-up approach adapting the structure of the system depicted in Fig. 1.3.

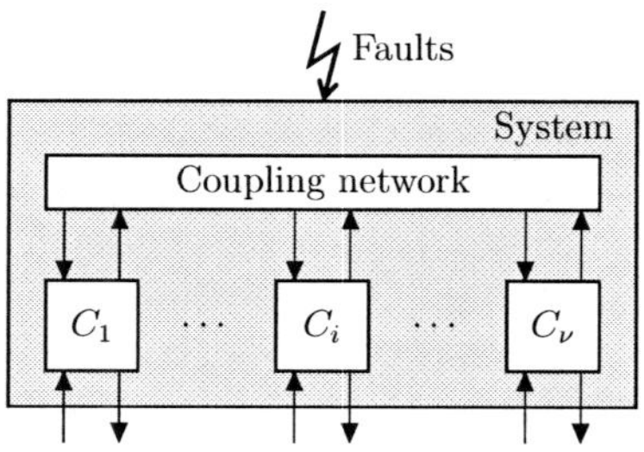

Figure 1.3: Network of coupled components C_i subject to faults

2. Analysis. The interaction in mechatronical systems may include the direct and instantaneous dependence of components on themselves, i.e. the problem of direct feedback may occur. Conditions have to be set up to test whether a given interconnection is well-posed or not. The discussion in Section 1.1 motivates the need of conditions for autonomy to allow for a partially decoupling of the network with regard to diagnosis. Finally, asynchronous state changes in the system have to be taken into account.

3. Partially coordinated diagnosis. A diagnostic algorithm must be designed for coupled mechatronical systems. The principle of consistency-based diagnosis introduced by [100] has been widely used in discrete-event literature. A consistency test between measurements of the system and the behavior of the model M is used to exclude faults in the case of inconsistency. This test is applied recursively to ensure online applicability. There are basically two lines of research which differ in the way of modeling faults: The diagnoser approach introduced by [106] considers faults as unobservable events whereas the multi-model approach introduced by [69] treats faults as an additional parameter of the behavioral relation. The diagnoser approach is not suitable in this framework because faults are assumed to severely change the behavior of the system.

The computational complexity prohibits the use of a single diagnostic unit D as shown in Fig. 1.1 for the whole system. Solving this problem is a wide area of research in diagnosis of discrete-event systems. The basic idea is to divide the diagnostic task into several subtasks. Five possible information structures are shown in Fig. 1.4: Centralized (Fig. 1.4(a)), decentralized (Fig. 1.4(b)), distributed (Fig. 1.4(c)), coordinated (Fig. 1.4(d)) and hierarchical (Fig. 1.4(e)). They will be discussed in Section 1.3.2 based on the diagnoser approach and in Section 1.3.3 using the multi-model approach. Most contributions in literature rely on the diagnoser approach, thus they appear to be not suitable for diagnosis of coupled mechatronical systems. There exists one approach using multi-models assuming synchronous interaction of the components [84]. The extension of this approach to asynchronous networks of I/O-automata will be given in this thesis. In the case of partial decoupled components, it will be examined in Chapter 6, if diagnosis can be carried out independently of the remaining network. A *partial coordination scheme* of the locally obtained diagnostic results will be analyzed in the case of coupled components to improve the diagnostic result of the system (Chapter 7).

1.3 Literature survey

1.3.1 Ways of modeling faults

Treating faults as unobservable events (Fig. 1.5) is investigated in Section 1.3.2 whereas the multi-model approach is presented in Section 1.3.3.

Faults as unobservable events. The most prominent way of modeling faults is considering them as unobservable events, i.e. events whose occurrence cannot be detected directly by sensors. A transition labeled with the fault event is added to the behavior of a standard automaton such that the model incorporates the fault-free and the faulty

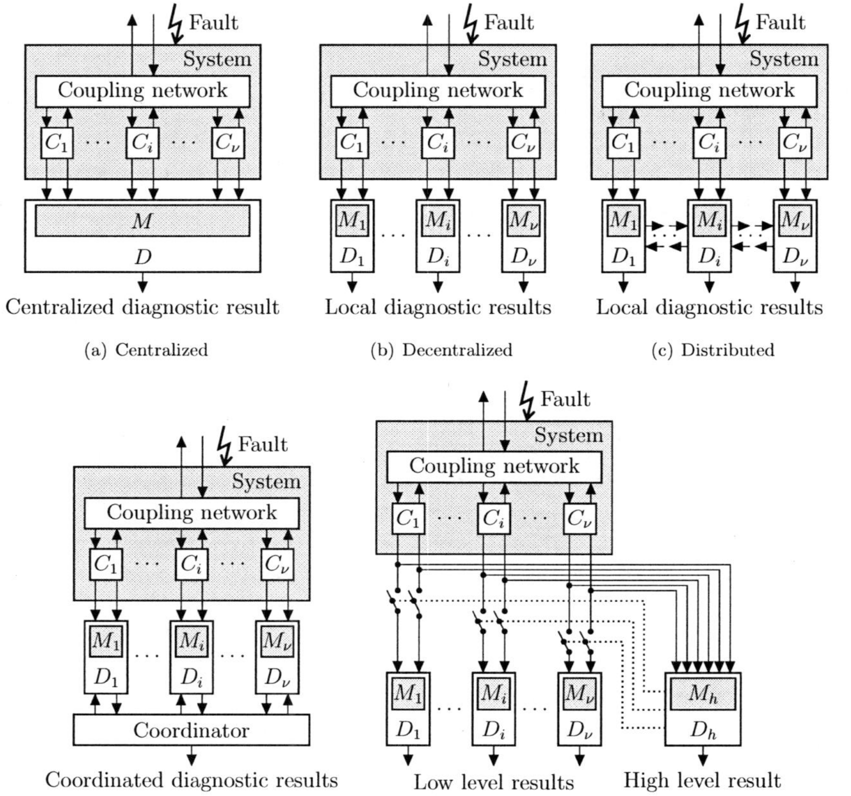

(a) Centralized (b) Decentralized (c) Distributed

(d) Coordinated (e) Hierarchical

Figure 1.4: Different information structures to diagnosis

behavior of the system [27, 106, 118] (the dotted line labeled with σ_f in Fig. 1.5). In this modeling approach, it is possible that the system may continue its operation after the occurrence of a fault. Hence, it is appropriate for faults that cause a distinct change in the system but do not necessarily stop the system. This idea is motivated from computer science [62] where the correct execution of a program may be interrupted by such a fault.

Faults as additional parameter of the behavioral relation. In contrast to the former approach, the fault is considered in this thesis as an additional parameter of the behavioral relation of an I/O-automaton. The motivation for this modeling philosophy results from the consideration of mechatronical systems as quantized systems. As described in [20, 113], the discrete-event model of systems subject to faults distinguishes with respect

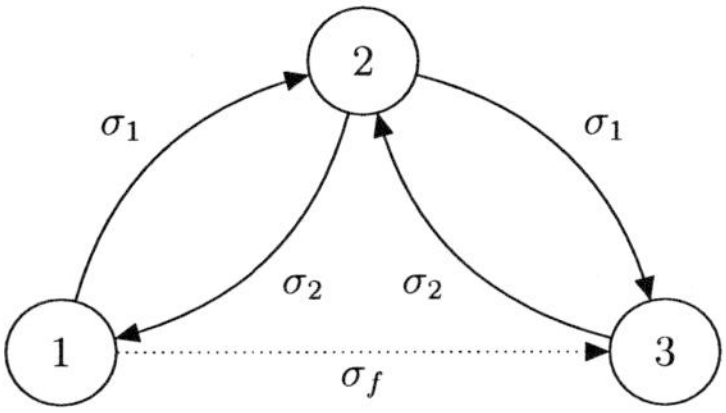

Figure 1.5: Faults modeled as unobservable events

to their behavioral relation, because the occurrence of a fault does not only move the system into a new state but, as long as the fault is present, the dynamical properties represented by the behavioral relation are different from the nominal ones (Fig. 3.1). Hence, separate models for the fault-free behavior and for each fault have to be set up as shown in Fig. 1.6. This modeling approach is used throughout this work but needs to be extended to cope with coupled mechatronical systems (Chapter 3).

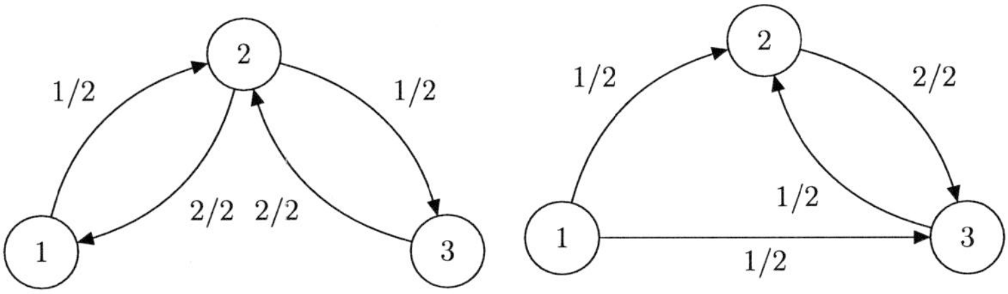

(a) Fault-free behavior (b) Faulty behavior

Figure 1.6: Faults modeled as additional parameter of the behavioral relation

1.3.2 Diagnoser approach

The diagnoser approach is the most prominent way to diagnosis of discrete-event systems and mostly related to model faults as unobservable events (Fig. 1.5). Thus, it is called the event-based framework of diagnosis in literature. The aim of diagnosis is to identify the occurrence of these fault events [61, 107, 108] or faulty traces that do not belong to a given specification language [95, 98] within a finite delay using model-based inferencing driven by the observed sequence of observable events. To achieve this, a diagnoser is used. It is a special type of standard automaton that is built by attaching fault labels (normal and faulty) to the system model and projecting the result onto the observable reach. Diagnosis terminates, if the diagnoser reaches a state, which describes the fault without any ambiguous cycle.

State-based approaches form another line of research using diagnosers. The state set of the system is partitioned according to conditions using a condition map. This conditions are related to normal or faulty operation of the system and referred to as failure states. Hence, faults are related to certain states and not to unobservable events. The aim of diagnosis in this setting is to estimate the conditions of the system. This is done using a different type of diagnoser which is a finite state machine taking the measured event sequence as input to generate the estimate of the failure state [48, 49, 99].

Enhancements of the monolithic approach. The aforementioned approaches suffer all from the *problem of state space explosion* because they relay on a monolithic model and a centralized diagnostic unit as depicted in Fig. 1.4(a). To partially reduce the complexity, a minimal set of sensors can be designed to fulfill the diagnostic task [43, 56, 64, 104, 124, 130]. Strategies on model reduction are analyzed in [49]. An approach based on the pre-computation of a reachability tree using the notion of basis markings and justification is presented in [24]. In [115] binary decision diagrams are introduced to improve space efficiency in storing the model and the diagnoser respectively. These strategies result in a more efficient representation of the monolithic model but do not solve the problem of state space explosion.

Hierarchical approaches are based on the abstraction of information to form layers in the model with different accuracy. Diagnosis is performed starting from the top layer description of the system to isolate faulty components (Fig. 1.4(e)). An improvement strategy is introduced using lower level models which contain more information of the component [81, 90, 119]. The associated lower level diagnostic units are used to identify the fault, if the higher level diagnostic result assumes these components as faulty. Different algorithms may be used at each level. On a higher level, a monolithic description may be suitable for diagnosis whereas this may be impossible on a lower level due to the interaction between involved components. Defining a hierarchy in the system description does not reduce the computational complexity in general. Hence, this approach will not be considered here.

Reduction of the computational complexity. Considerations on further reduction of the complexity are an important issue in current research on diagnosis of discrete-event systems. The only way to solve this problem is to map the structure of the system onto the diagnostic system, i.e. to work with a modular structure in modeling the system [22] and a division of the diagnostic task based on this structure. The basic approaches [33, 34, 57, 118] are top down approaches to design a modular diagnostic system, i.e. the global model is assumed to be given. Local diagnosers with a corresponding communication mechanism are built from the corresponding given centralized diagnoser. A trade-off must be made depending on the wish to ensure equivalence of the modular with the centrally obtained result and the amount of data that needs to be communicated.

The fewest amount of data is communicated in the *decentralized* set-up where local diagnosers are assigned to the components. They act independently of each other as shown in Fig. 1.4(b). The diagnostic result is reported to a centralized component. It degrades in general due to the neglected interaction between the components [28, 34, 95]. Even though the diagnosers operate in a modular set-up, the computational complexity is still high for the check of diagnosability because it is a property of the system which cannot be solved solely based on the component models. Solutions to overcome this problem based on efficient algorithms are given in [54, 83, 134, 138]. A different approach based on verifiers is given in [91, 129, 135]. The question of dealing with robustness against the total failing of one or more local diagnoser is answered in [18].

In [30, 31, 42] a modular decomposition is introduced to extend the locally measurable event set to reach global diagnosability, i.e. to ensure the equality of the decentralized and the centralized diagnostic approach without communication between the local diagnosers but restricting the system structure. In [29] this problem is solved by defining state- and transition independency conditions which are assumed to hold for the approach presented in [93, 109].

The decentrally obtained diagnostic result is enhanced in the *distributed* set-up by exchanging data among the local diagnosers (Fig. 1.4(c)). The same result as in the centralized approach can be achieved. An algorithm based on rounds in communication is introduced in [118, 120]. It allows only for communication between adjacent diagnosers. The algorithm is guaranteed to terminate from a local point of view, only if the system under consideration has a tree structure. In [58] a justification tree-based approach is introduced to design an accurate diagnoser revealing the maximum number of couplings in the system. Another approach to reduce the number of communication channels is presented in [21, 102] by allowing only an asymmetric flow of information from one local diagnoser to another one. A defacto decoupling of the system is given, if the interaction between the components is assumed to be observable as in [14, 44, 45] where communication may take place between all local diagnosers.

The distributed set-up is called *coordinated*, if the local diagnosers are not allowed to communicate among each other but via a central component called coordinator as shown in Fig. 1.4(d) [15, 33, 34]. The advantage of this setting compared to the distributed one is its ability to cope with a general interaction structure because all locally obtained diagnostic results are collected in the coordinator to obtain the global diagnostic result. The drawback of the cited literature is the assumption that the interaction between the components is measurable which results in a defacto decoupling of the components.

Unreliable communication. A *reliable communication* between the local diagnosers has been assumed in all previously discussed modular settings in the sense that the global ordering of locally measured events is preserved under communication. This assumption is relaxed in [35, 97] by introducing communication delays. The order of emission of events is not equal to the order of reception of them. In addition, the authors of [92] have found a way to explicitly describe the interaction between the system and its environment and among the components respectively. It has been shown in [96] that a distributed setting with unbounded communication delays results defacto in the decentralized setting. A different approach has been introduced by [19, 38, 47] assuming asynchronous communication between subsystems and the centralized diagnostic unit. To account for the resulting communication delays, the concept of net unfolding is used.

Further reduction of complexity. Given the results of distributed or coordinated diagnosis, basically two different types of further improvements have been considered in literature. First, concepts of minimizing the amount of exchanged data in a given communication structure have been studied in [67, 128]. Second, the concepts of efficiently representing the model discussed in conjunction with the monolithic approach can be extended to the modular setting. In [94], causal networks are used to reduce the complexity of the model. The concept of basis markings and justification based on a pre-computed reachability tree is modularized in [25]. The extension of binary decision diagrams to cope with the modular structure is given in [116].

1.3.3 Multi-model approach

The multi-model approach is mostly used in the context of quantized systems [20, 39, 69, 113, 136] where faults are assumed to be unknown inputs of the system. They are modeled as additional parameters of the behavioral relation (Fig. 1.6). The aggregation of all models, each describing the system subject a fault including the fault-free case, yields the system model [16]. Events are separated in this context usually into input and output events that occur simultaneously to cover the cause-and-effect-chains of the system. Some contributions use multi-models in conjunction with a single event set containing input and output events [3, 65]. The aim of diagnosis is in both cases to estimate the (discrete) value of the fault signal based on the measurements of the system using an observation algorithm and the system model. The consistency test is applied to all modeled faults recursively to ensure online applicability. A fault will be excluded from the set of possible faults, if the measured system behavior is inconsistent with the model of this fault. Testing the diagnosability of I/O-automata is, as opposed to the diagnoser approach, a fairly new branch of research [74, 75].

The multi-model approach has the advantage that the system model is directly used for diagnosis such that no diagnoser needs to be pre-computed as in the diagnoser approach. Besides this, the separation between the system model and the diagnostic algorithm is given.

Enhancements of the monolithic approach. Solving the *problem of state space explosion* is also an active field of research in this context. One idea is to use an *hierarchical approach* [114, 122, 121] as discussed in Section 1.3.2 with the associated problems. An extension to this approach is to split up the diagnostic task into on-board fault detection and off-board fault identification due to computation restrictions of the on-board system [41]. This set-up is referred to as *remote diagnosis*. The size of the model can be further reduced as shown in [40]. The drawback of this approach is the need to communicate measurements via mostly unreliable communication channels, if a fault has been detected by the on-board system. A method that is capable to cope with these effects has been proposed in [110, 111]. Even though the computational complexity of the on-board diagnostic unit is reduced, the problem may still exist in the off-board unit.

Reduction of the computational complexity. Further considerations on complexity reduction are also an important issue in diagnosis using the multi-model approach. The most obvious solution uses a centralized setting with an algorithm for online composition of that part of the global model that is needed based on the actual measurements [78]. The drawback of this approach is the loss of the structure in the model. In [77], conditions are presented to decouple the system into subsystems that can be diagnosed separately.

A *decentralized* information structure as shown in Fig. 1.4(b) is used in [85]. The diagnostic result degrades because communication between local diagnosers does not exist to estimate the value of immeasurable interactions. The *distributed* framework depicted in Fig. 1.4(c) has not been considered in modeling faults as an additional parameter of the behavioral relation and is not further investigated.

The complete reduction of the computational complexity is given in [84] for *synchronous networks of I/O-automata* where a clock signal is used to periodically update the value of the discrete input and output signals and the immeasurable internal couplings respectively. This approach may result in direct feedback, if the output of a component depends directly and instantaneously through a signal path within the network on itself. Correspondingly, the interconnection may not be well-posed [86]. Two different *coordination schemes* are analyzed for well-posed systems: Unidirectional in [88] and bidirectional in [87]. Only the latter structure (Fig. 1.4(d)) guarantees the equivalence of the diagnostic result with the centrally obtained diagnostic result at all but the highest amount of communicated data is needed in this setting.

1.3.4 Literature evaluation

Diagnosis of discrete-event systems is an active area of research. It can be separated in the *diagnoser approach* discussed in Section 1.3.2 and the *multi-model approach* considered in Section 1.3.3. The complexity problem, i.e. the exponential growth of the state space due to a linearly increasing number of components, is common to both methods. Several ways of reducing the complexity based on the information structures shown in Fig. 1.4 have been developed. A decision for the deployed setting depends on the trade-off between the desired accuracy of the diagnostic result and the allowed amount of exchanged data. A short evaluation of the aforementioned approaches with respect to methods for the complexity reduction of discrete-event modeling and diagnosis of coupled mechatronical systems with partially autonomous behavior will be given according to the three tasks stated in Section 1.2.

1. Component-oriented modeling. The model must have the ability to explicitly represent the cause-and-effect-chains of the system, to cope with asynchrony and to treat faults as an additional parameter of the behavioral relation. The basic decision must be taken between standard automata and I/O-automata, which allow both for the desired fault modeling philosophy. Even though, this philosophy is not commonly used within the first setting [27, 36, 63, 70]. Coupled standard automata allow for an asynchronous motion of some components but cover the interaction between components implicitly based on the event names, whereas synchronous networks of I/O-automata [84, 113] explicitly express the cause-and-effect-chains of the system using signals to model the interrelation with the drawback of allowing only for synchronous behavior of the components and modeling the interaction between the components based on signal names. A modeling method combining the advantages of both approaches has not yet been developed in literature. Hence, it is developed in this thesis for asynchronous networks of I/O-automata representing coupled mechatronical systems.

2. Analysis. Coupling of component models can result in general in loops. This interconnection may not always be *well-posed.* This phenomenon is known form synchronous networks of I/O-automata and has been solved in [84, 86]. Their criterion has to be extended to the new modeling formalism. Conditions for partial autonomy of several components have not been found in literature whereas asynchronous motion can be modeled by introducing the empty symbol ε. This approach is known from standard automata modeling [27, 92] and has been extended to networks of I/O-automata without coupling signals in [70]. The main idea of the empty symbol is to explicitly describe the situation where one component moves and another one does "nothing", i.e. does neither change its state nor generates an event. This concept can be used for coupled mechatronical systems to model the non-interaction of those operating points to fulfill the design task.

3. Partially coordinated diagnosis. The multi-model approach is used for diagnostic purposes because the diagnoser approach can only handle faults modeled as unobservable events. The extension of the principle of consistency-based diagnosis to the multi-model approach in conjunction with an I/O-description of the system is explained in Chapter 2. The idea of partial composition of the behavioral relation considered in [78] can be used for *centralized diagnosis* to serve as a reference for the analysis of modular approaches because it refuses the system structure. The *decentralized* and the *coordinated* set-ups introduced in [84] reduce the complexity problem based on synchronous networks of I/O-automata with the drawbacks described above. The ideas behind the concepts will be used here for diagnosis of asynchronous networks of I/O-automata: The decentralized framework provides the best diagnostic result in the case of autonomy and the coordination scheme gains equality in the case of interaction between the components. The combination of both approaches results in a partially coordinated diagnostic framework.

1.4 Main results of this thesis

The main results of this thesis are methods for modeling and diagnosis of coupled mechatronical systems with partial autonomous behavior which reduce the computational complexity of the discrete-event modeling and of the diagnosis for this class of systems. They can be split up according to the tasks stated in Section 1.2.

1. Component-oriented modeling. The following results concerning the modeling of asynchronous discrete-event systems have been achieved:

- *Asynchronous networks of I/O-automata* are developed to cope with the modeling of partial autonomy in the operation of mechatronical system. The preliminary study [2] is extended to the approach presented in Chapter 3. It has been published in [5]. I/O-automata are used as component models. The components' measurable inputs and outputs are modeled as control signals. The interaction between the components is described via the newly introduced coupling signals. They are linked via an coupling block to one another. Hence, the cause-and-effect-chains of the system are copied directly into the model. Changes in the system can be easily adapted in the model due to its modular structure. The composition operators to build an equivalent single I/O-automaton are given by eqns. (3.78) and (3.79).
- Faults are modeled as an additional parameter of the behavioral relation [20] due to the assumption that they severely change the dynamical properties of the system. As a result, multi-models are used to describe the components.
- The new modeling formalism is extended in Section 3.6 to cope with asynchronous state transitions by introducing the empty symbol ε to model the non-interaction between components in those operating points where the design task is fulfilled by these components separately [5, 72]. Consequently, the sequences of all local measurements are of equal length and the behavior of all components is synchronized upon the occurrence of events in the system.
- The new modeling concept is based on the idea of parallel composition known from standard automata [27, 70] where the interactions among the components are implicitly given by the name of the events. It is shown in Section 3.8 and in [5] that the new modeling formalism is applicable for at least the same class of asynchronous discrete-event systems as the known modeling formalism while having certain advantages. The formal proof of equivalence can be found in [73].

2. Analysis. The analysis of asynchronous networks of I/O-automata comprises the following issues:

- Coupling of component models can result, in general, in loops, where the behavior of a component depends directly and instantaneously on itself via a signal path within the network. This interconnection may not always be *well-posed.* This phenomenon is known from synchronous networks of I/O-automata discussed in [84, 86]. Their criterion has been extended in the diploma thesis [11] and published in [6] on an algorithmic base for deterministic asynchronous networks of I/O-automata. The corresponding mathematical description is introduced in Section 4.1 and extended to the nondeterministic case. It is shown, that the equivalent single I/O-automaton obtained by composition of the deterministic asynchronous network of I/O-automata can either be deterministic (Theorem 1), weakly deterministic (Theorem 2) or nondeterministic (Theorem 3) in the case of nondeterministic asynchronous networks of I/O-automata.

- Two types of autonomy are introduced in this thesis. Conditions for *structural autonomy* are derived in Section 4.2.1 based on the search for sets of strongly connected nodes in the corresponding directed graph of the system which are coupled in neither way. It is assumed in the sequel that the system under consideration cannot be separated into independent subsystems such that no simplifications are possible according to diagnosis. *State-dependent autonomy* is investigated in Section 4.2.2 in terms of the behavioral relation. It is the foundation for simplifications in diagnosis of discrete-event systems. It has been published in [72] on a reduced version of the new modeling formalism which does not allow for algebraic loops.

- Two algorithms for the *simulation* of the behavior of asynchronous networks of I/O-automata are presented in Section 4.3. The first one uses the whole equivalent I/O-automaton which has been pre-computed by composition of the network, whereas the second one reduces the model size by applying the composition rule only to the part of the behavioral relation concerning the actual state and inputs. The equality of their results is stated in Theorem 4. The latter approach will be referred to as *online composition* in the sequel.

3. Partially coordinated diagnosis. The main results of this thesis concern methods for the diagnosis of asynchronous discrete-event systems:

- As an intermediate step towards the diagnostic methods presented in this work, the equality of centralized and decentralized diagnosis has been proven, if faults are modeled as an additional parameter of the behavioral relation of standard automata which are coupled by parallel composition. This approach has been published in [1, 3] and applied in [4] to the modular diagnosis of the VDM. As it does not rely on I/O-automata, it will not be discussed in this work.

- A method for *state observation* of discrete-event systems is analyzed in Section 5.1. Even though it builds the basis for the diagnostic algorithms presented in this thesis, it is not the focus of this work. For a better understanding of the diagnostic methods, state observation is explained in the centralized set-up only. Theorem 5 states that the observation result is identical to the ideal observation result: $\mathcal{Z}^c(k_e) = \mathcal{Z}^\star(k_e)$.

- The method for state observation is extended in Section 5.2 to centralized diagnosis of a single I/O-automaton. The presented diagnostic algorithm yields the ideal diagnostic result: $\mathcal{F}^{dc}(k_e) = \mathcal{F}^\star(k_e)$ but does not solve the complexity issue. Online composition of the behavioral relation using a centralized diagnostic unit known from [78] is extended to cope with asynchronous networks of I/O-automata (Section 5.3). This approach yields the ideal diagnostic result (Theorem 7) but reduction of the computational complexity is rather small. Hence, it serves as a reference in here.

- To reduce the computational complexity further, the diagnostic task is solved in a *decentralized* way by extending [85] to the new modeling concept next (Fig. 1.4(b)). It is online applicable and needs the lowest amount of storage capacity and computing time of all presented approaches. The decentralized diagnostic method is introduced in Chapter 6 for the general case allowing for direct dependencies among the components, a simplified version prohibiting these dependencies can be found in [72]. An extended version of the centralized diagnostic algorithm is used in the local diagnostic units to compute locally the ideal diagnostic result: $\mathcal{F}_i^{dc}(k_e) = \mathcal{F}_i^\star(k_e)$ (Theorem 8). All locally obtained diagnostic results which are synchronized upon the occurrence of events are combined in the so-called merger to get the decentrally obtained diagnostic result $\mathcal{F}^{dc}(k_e)$. It is identical to the ideal diagnostic result $\mathcal{F}^\star(k_e)$ in the case of state-dependent autonomy for well-posed networks (Theorem 11). A lack of soundness arises in the general case (Theorem 10) due to the negligence of the interrelation between the components such that the decentrally obtained diagnostic result degrades: $\mathcal{F}^{dc}(k_e) \supseteq \mathcal{F}^\star(k_e)$ (Theorem 9). Hence, the best diagnostic result is only obtainable in the case of state-dependent autonomy based on the most restrictive information structure.

- To overcome the soundness issue in generell, a method for *partially coordinated diagnosis* is introduced in Chapter 7. It extends the bidirectional coordination scheme introduced in [87, 88] to asynchronous networks of I/O-automata to obtain the ideal diagnostic result based on the structured model. The locally obtained sets of possible pairs of interconnection inputs, outputs and faults are calculated in the local diagnostic units and sent to a centralized coordinator. This coordinator removes contradictions from these sets to obtain the partially coordinated diagnostic result $\mathcal{F}^{pc}(k_e)$ which is always identical to the ideal diagnostic result: $\mathcal{F}^{pc}(k_e) = \mathcal{F}^\star(k_e)$ (Theorem 12). It is contravened to [84] that the coordinator must have a *memory* to store the partially coordinated diagnostic result for the next time step.

The following simplifications result from state-dependent autonomy: The local diagnostic units of those components being in state-dependent autonomy do not get back the result of coordination. Coordination must not take place if the coordinator has decided that all components are in state-dependent autonomy. Consequently, no local diagnostic unit will get back any information from the coordinator.
The presented method for partially coordinated diagnosis is online applicable. It reduces the computational complexity significantly but not as much as the decentralized set-up does due to the extra computing time needed for coordination. The developed diagnostic algorithm is applicable to synchronous networks of I/O-automata due to the use of the empty symbol.

4. Application example. Two application examples are presented in Chapter 8. The decentralized and the partially coordinated diagnostic algorithm are applied to a process plant (Section 8.1). The shown results relay on the Study thesis [9]. The simulation of the production facility known from [70] is considered in Section 8.2. These results have been achieved during the Diploma thesis [11].

5. Implementation. A MATLAB toolbox has been developed for modeling, analysis and diagnosis of asynchronous networks of I/O-automata. The development started during the internship [10] using coupled standard automata. It has been extended in the Diploma thesis [11] to cope with deterministic asynchronous networks of I/O-automata introduced in Sections 3.2 to 3.6 and the analysis described in Section 4.1. Extensions to the nondeterministic case started during the internship [7] and have been finished during the internships [8] to handle faults as an additional parameter. Besides designing a graphical user interface, the method for centralized diagnosis described in Chapter 5 has been implemented during the latter works. The finial step towards partially coordinated diagnosis introduced in Chapter 7 has been made in the study thesis [9].

1.5 Document structure

Chapter 2 gives the basic concept of the principle of consistency-based diagnosis. This principle is extended to discrete-event systems using the multi-model approach based on I/O-automata where faults are considered as an additional parameter of the behavioral relation.

Chapter 3 introduces the new approach for component-oriented modeling of asynchronous discrete-event systems. It starts with an explanation of the basic concepts of qualitative modeling and composite discrete-event systems. Each component is modeled independently of the remaining components by a special kind of I/O-automaton. The interconnection of the components is given by a general coupling model. The notion of asynchrony is realized by usage of the empty symbol. The composition of the resulting asynchronous network of I/O-automata to obtain an equivalent single I/O-automaton is described as well as the usage of a model library with the accompanying problem of unconnected coupling signals. By using the parallel composition rule known from standard automata theory as an example, it is finally shown that the new model is applicable for at least the same class of asynchronous discrete-event systems as the known modeling formalisms.

Chapter 4 gives solutions to the feedback problem arising from loops within the network. Conditions for structural and state-dependent autonomy in the system are investigated. Two algorithms for the simulation of the behavior of nondeterministic asynchronous networks of I/O-automata are given as well as considerations on the complexity of the model.

Chapter 5 is concerned with the centralized diagnosis of nondeterministic processes. A method for state observation is investigated for the centralized set-up only. It is extended to diagnosis of a single I/O-automaton and asynchronous networks of I/O-automata.

Chapter 6 gives solutions to the decentralization of the diagnostic task. The diagnostic result is analyzed for the general case allowing for interaction between the components and the simplifications resulting from state-dependent autonomy. The complexity of the diagnostic algorithm is investigated.

Chapter 7 deals with the development of the partially coordinated diagnostic method. The complexity of the diagnostic algorithm is analyzed. An example to compare all diagnostic approaches presented in this thesis is given. The chapter closes with an evaluation of the presented diagnostic approaches.

Chapter 8 gives two examples to demonstrate the application of the approaches presented in this thesis: The decentralized and partially coordinated diagnosis of a process plant as well as the simulation of a production facility using online composition.

Chapter 9 summarizes and concludes this thesis. It gives an outlook on open problems and possible research topics related to this work.

Chapter 2

Consistency-based diagnosis

The diagnostic methods presented in this thesis relay on the principle of consistency-based diagnosis which is explained in Section 2.1. The application to discrete-event systems is given in Section 2.2 using the multi-model approach based on I/O-automata.

2.1 Basic concepts

The principle of *consistency-based diagnosis* is a model-based approach using the structure shown in Fig. 1.1. It has been introduced in [100] and widely used in qualitative reasoning which belongs to the area of artificial intelligence [13, 52, 53, 76, 101]. The model used for diagnosis consists of a *system description* (SD) based on logic formulas, which describe the behavior of these components, and a set of *components* ($COMPS$) that include propositions about faults in the components. It is used to check for conflicts $\perp$ with the *observations* (OBS) which consist of a set of formulas describing the measurements of the system. The aim of diagnosis is to infer about the fault from altering the formula in $COMPS$ such that the conflict does not exist any longer. The qualitative reasoning $\models$ is given by the relation

$$COMPS \cup SD \cup OBS \not\models \perp . \tag{2.1}$$

This procedure results in a set of *fault candidates* which includes the true fault, if it has been considered during modeling. The best known implementation is the General Diagnostic Engine (GDE) [32]. It is based on the decomposition of the system into smaller components that are interconnected by qualitative signals.

The most burdensome drawback of this logic-based approach is its restriction on static systems, i.e. changes in the dynamics of the system cannot be diagnosed. To overcome this problem, models that are able to cope with the dynamical properties of the system have to be considered. The fundamental description of the application of the principle of consistency-based diagnosis to discrete-event systems is given in the following.

2.2 Application to discrete-event systems

The sequences of input events $\boldsymbol{v}(k)$ that trigger the system and the sequence of output events $\boldsymbol{w}(k)$ that are generated by the system (Fig. 2.1) are abbreviated by

$$\boldsymbol{V}(0\dots k_e) = (\boldsymbol{v}(0), \boldsymbol{v}(1), \dots, \boldsymbol{v}(k), \dots, \boldsymbol{v}(k_e)) \text{ and}$$
$$\boldsymbol{W}(0\dots k_e) = (\boldsymbol{w}(0), \boldsymbol{w}(1), \dots, \boldsymbol{w}(k), \dots, \boldsymbol{w}(k_e)).$$

The integer number $k \in \mathbb{N}$ counts the number of elapsed events. The number of the last event of the event sequence is denoted by k_e.

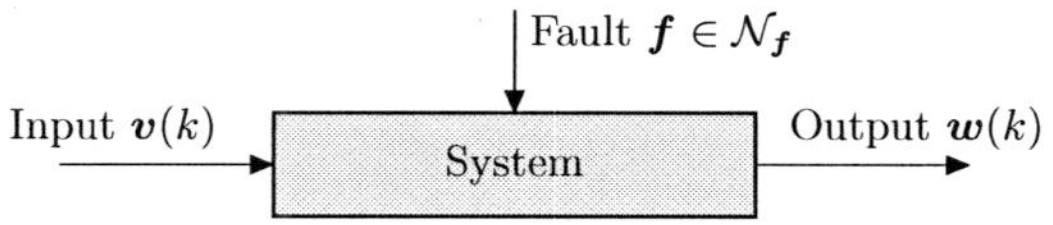

Figure 2.1: Discrete-event input $\boldsymbol{v}(k)$ and output $\boldsymbol{w}(k)$ of a system

The set of considered faults

$$\mathcal{N}_{\boldsymbol{f}} = \{0, 1, \dots, S\} \tag{2.2}$$

consists of the faults the system may be subject to. If *all* possible faults have been regarded in the modeling process, the so-called *closed world assumption* holds. The faultless case is included in the set of faults as $\boldsymbol{f} = 0$ to obtain a uniform notation. Without loss of generality, *constant* [1] faults are assumed in this thesis, i.e. the fault has occurred prior to diagnosis and does not change until diagnosis is finished.

In the context of discrete-event systems, the measurement $\boldsymbol{B}(k_e)$ of the system

$$\boldsymbol{B}(k_e) = (\boldsymbol{V}(0\dots k_e), \boldsymbol{W}(0\dots k_e)) \tag{2.3}$$

is given as an I/O-sequence of length k_e. The set of all I/O-sequences of arbitrary length which can be generated by the system defines the behavior $\mathcal{B}_S$ of the system

$$\mathcal{B}_S \subseteq \mathcal{N}_{\boldsymbol{v}}^* \times \mathcal{N}_{\boldsymbol{w}}^*, \tag{2.4}$$

where $\mathcal{N}_{\boldsymbol{v}}^* \times \mathcal{N}_{\boldsymbol{w}}^*$ denotes the Kleene closure [51] of the set of input symbols $\mathcal{N}_{\boldsymbol{v}}$ and the set of output symbols $\mathcal{N}_{\boldsymbol{w}}$.

The behavior of the system under the influence of a given fault $\boldsymbol{f} \in \mathcal{N}_{\boldsymbol{f}}$ is described in a *multi-model* as proposed in Section 1.3.3 where each fault is assigned to its own model $\mathcal{M}(\boldsymbol{f})$. The consistency test defined in eqn. (2.1) is used to detect and identify the fault that influences the system by comparing the measurement $\boldsymbol{B}(k_e)$ to the behavior $\mathcal{B}_{\mathcal{M}}(\boldsymbol{f})$ of the model that includes all I/O-sequences modeled by $\mathcal{M}(\boldsymbol{f})$. This concept is shown in Fig. 2.2.

[1] Approaches to handle fault dynamics can be found [20, 84, 113]. The special case of repeatable faults is considered in [55, 137], whereas [26, 66] consider sensor failures.

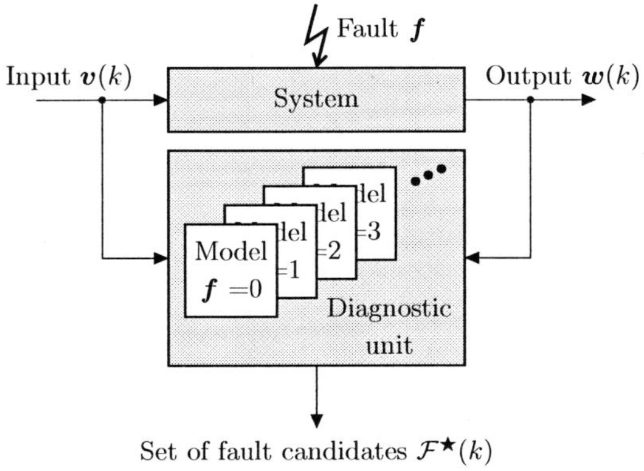

Figure 2.2: Basic concept of diagnosis

Definition 1. *(Consistency) The measurement $\boldsymbol{B}(k_e)$ is said to be* consistent *with the model $\mathcal{M}(\boldsymbol{f})$, if*

$$\boldsymbol{B}(k_e) \in \mathcal{B}_{\mathcal{M}}(\boldsymbol{f}) \tag{2.5}$$

holds. ◇

Definition 2. *(Fault candidate) A fault $\boldsymbol{f}$ is a fault candidate, if the model $\mathcal{M}(\boldsymbol{f})$ of the system subject to the fault $\boldsymbol{f}$ is consistent with the measurement $\boldsymbol{B}(k_e)$.* ◇

The set of fault candidates $\mathcal{F}^{\star}(k_e) \subseteq \mathcal{N}_{\boldsymbol{f}}$ contains all faults for which the consistency test (2.5) holds up to time k_e. It is given as

$$\mathcal{F}^{\star}(k_e) := \{\boldsymbol{f} | \boldsymbol{B}(k_e) \in \mathcal{B}_{\mathcal{M}}(\boldsymbol{f})\} \tag{2.6}$$

and referred to as the *ideal diagnostic result.* Eqn. (2.6) can be rewritten as

$$\boldsymbol{f} \in \mathcal{F}^{\star}(k_e) \Leftrightarrow \boldsymbol{B}(k_e) \in \mathcal{B}_{\mathcal{M}}(\boldsymbol{f}). \tag{2.7}$$

Usually, the behaviors $\mathcal{B}_{\mathcal{M}}(\boldsymbol{f})$ of system models subject to the fault $\boldsymbol{f}$ are over-lapping for different faults, i.e. there exist I/O-sequences that can be generated by more than one model. By this, $\boldsymbol{B}(k_e)$ might be consistent with several models. Hence, $\mathcal{F}^{\star}(k_e)$ contains additional faults besides the fault that actually influences the system. This consistency test is used in the remainder of this thesis to develop a diagnostic algorithm that yields a diagnostic result as close as possible to the ideal result $\mathcal{F}^{\star}(k_e)$.

Due to the fact that $\mathcal{F}^{\star}(k_e)$ may contain more faults than the actual one, its elements are called *fault candidates.* It is assured that no fault $\boldsymbol{f}$ is missing in $\mathcal{F}^{\star}(k_e)$ because the consistency test only excludes those faults that do not fulfill condition (2.5). Thus, a wrong exclusion of a fault is impossible. The diagnostic result is said to be *complete* and *sound.* The following Definitions 3 and 4 are originated from [84].

Definition 3. *(Completeness) A diagnostic result $\mathcal{F}(k_e) \subseteq \mathcal{N}_{\boldsymbol{f}}$ is called* complete, *if all fault candidates are contained in the set $\mathcal{F}(k_e)$:*

$$\mathcal{F}(k_e) \supseteq \mathcal{F}^{\bigstar}(k_e). \tag{2.8}$$

◊

The following relation holds according to Definition 3:

$$\boldsymbol{f} \in \mathcal{F}(k_e) \Leftarrow \boldsymbol{B}(k_e) \in \mathcal{B}_{\mathcal{M}}(\boldsymbol{f}). \tag{2.9}$$

Definition 4. *(Soundness) A diagnostic result $\mathcal{F}(k_e) \subseteq \mathcal{N}_{\boldsymbol{f}}$ is called* sound, *if all faults $\boldsymbol{f}$ in the set $\mathcal{F}(k_e)$ are fault candidates:*

$$\mathcal{F}(k_e) \subseteq \mathcal{F}^{\bigstar}(k_e). \tag{2.10}$$

◊

This definition results in the following relation

$$\boldsymbol{f} \in \mathcal{F}(k_e) \Rightarrow \boldsymbol{B}(k_e) \in \mathcal{B}_{\mathcal{M}}(\boldsymbol{f}). \tag{2.11}$$

A complete diagnostic result that is not sound may contain additional faults that are no fault candidates. These additional faults

$$\boldsymbol{f} \in \mathcal{F}^{spur}(k_e) = \mathcal{F}(k_e) \backslash \mathcal{F}^{\bigstar}(k_e) \tag{2.12}$$

are called *spurious solutions* and may cause false alarm [84]. Opposed to this, the relation $\mathcal{F}(k_e) \subseteq \mathcal{F}^{\bigstar}(k_e)$ indicates a sound diagnostic result that is not complete. Hence, not all fault candidates may be included in $\mathcal{F}(k_e)$. Due to a lack of completeness a broken system may be diagnosed as faultless. Based on this annotations, the completeness property is *required* for all diagnostic methods presented here.

With this convention a fault is *detected*, if the faultless case has been excluded from the diagnostic result:

$$0 \notin \mathcal{F}(k_e).$$

A fault $\boldsymbol{f} = i$ has been *identified*, if it is the only element in the diagnostic result:

$$\{i\} = \mathcal{F}(k_e).$$

A system is considered to operate faultless, if $\boldsymbol{f} = 0$ has been identified. This statement assumes that the closed world assumption holds.

The system will be described by I/O-automata in this work. The models $\mathcal{M}(\boldsymbol{f})$ are joined to one model $\mathcal{M}$ where the unknown fault signal $\boldsymbol{f}$ is considered as an additional parameter of the behavioral relation (Fig. 1.6). In addition, it is assumed that the behavior of the model is *identical* to the system:

$$\mathcal{B}_{\mathcal{M}} = \mathcal{B}_{\mathcal{S}}. \tag{2.13}$$

Hence, the model is complete and sound:

$$\boldsymbol{B}(k_e) \in \mathcal{B}_{\mathcal{M}} \Leftrightarrow \boldsymbol{B}(k_e) \in \mathcal{B}_{\mathcal{S}}. \tag{2.14}$$

In general, this assumption may not be realistic. Concepts for dealing with modeling errors can be found in [68, 123]. To ensure that no system behavior is missed, (2.13) can be relaxed to

$$\mathcal{B}_{\mathcal{M}} \supseteq \mathcal{B}_{\mathcal{S}}. \tag{2.15}$$

Then, the behavior of the model may contain I/O-sequences that do not belong to the system behavior. The modeling process is not investigated in detail in this work. Different modeling approaches can be found in [30, 39, 84, 105, 113, 121, 136].

Demands on the diagnostic method. The aforementioned statements can be summarized to the following demands on the diagnostic methods presented in this thesis:

- The closed world assumption, i.e. *all* possible faults have been regarded in the modeling process, is assumed to hold in this thesis to allow for the above mentioned statements about fault detection and isolation.
- All diagnostic approaches are required to be complete. Hence, a broken system can never be diagnosed as faultless.
- The model of the system is complete and sound, i.e. eqn. (2.13) holds. Hence, the consistency of the model with the measurements ensures the consistency of the system with the measurements.

Chapter 3

A new modeling formalism for asynchronous interconnected discrete-event systems

A new approach for component-oriented modeling of asynchronous discrete-event systems is presented in this chapter. Basic concepts of qualitative modeling are explained in Section 3.1. The composite discrete-event system is described in Section 3.2. Each component is modeled independently of the remaining components by a special kind of I/O-automaton that is introduced in Section 3.3. The interconnection of the components is given in Section 3.4 by a general coupling model. The resulting asynchronous network of I/O-automata and its composition to an equivalent single I/O-automaton are considered in Section 3.5. The notion of asynchrony is investigated in Section 3.6. The usage of a model library with the accompanying problem of unconnected coupling signals is described in Section 3.7. By the use of the parallel composition rule known from standard automata theory as an example, it is shown in Section 3.8 that the new model is applicable for at least the same class of asynchronous discrete-event systems as the known modeling formalisms.

3.1 Basic concepts

In qualitative modeling, the information about the system and its behavior is abstracted from a continuous to a symbolic description. Each time the continuous signal crosses a given bound, an event is generated that indicates the corresponding change in the symbolic value of the signal [27, 70]. This concept can be explained using Fig. 3.1 where the depicted signals may be seen as the output of a system. If the continuous signal, shown as a solid line in the topmost diagram, crosses, for example, the topmost dotted line, the event σ_{nh} pictured as an upright arrow is generated. This event indicates that the signal value

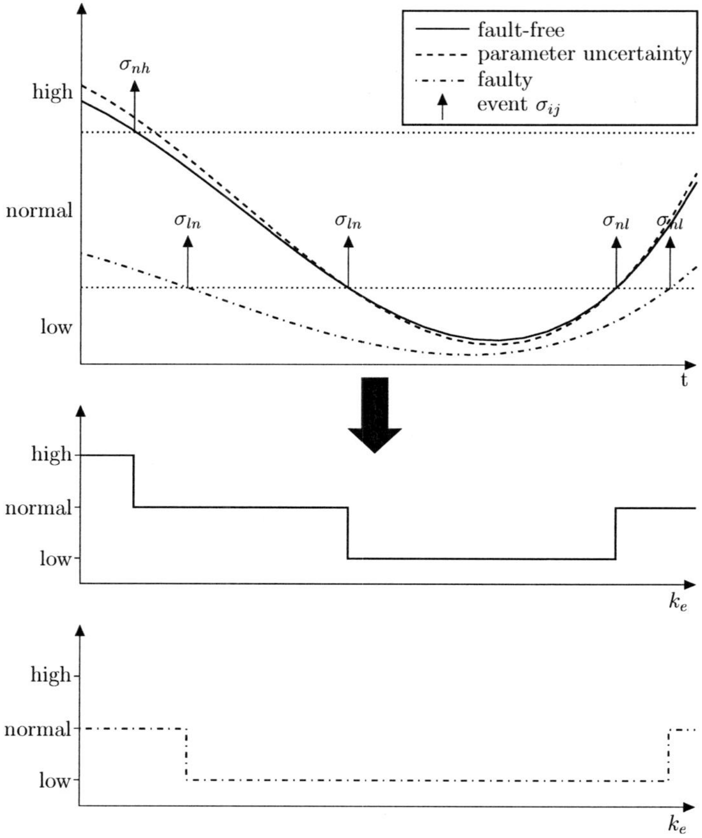

Figure 3.1: Abstraction of a qualitative signal and the effect of uncertainties versus faults

changes from high to normal as shown in the middle diagram. The resulting symbolic description of the continuous signal is given as the sequence (high, normal, low, normal) where signal changes are indicated by the events that are generated when the continuous signal crosses the corresponding bound.

The important aspects of qualitative modeling for diagnostic purposes can be explained as follows. First, the model is robust against parameter uncertainties. Consider the dashed line in the topmost diagram of Fig. 3.1: Parameter uncertainties or changes in the environmental conditions may lead to slight deviations of the continuous signal but do not affect the symbolic sequence. In contrast to this, a fault may result in severe changes of the dynamics as shown by the dashed and dotted line. This kind of change affects the qualitative behavior in the sense that the symbolic sequence (normal, low, normal) is at hand as shown in the undermost diagram. The symbolic signal changes due to the generation of the events σ_{ln} and σ_{nl}. Hence, taking a look at the symbolic sequence is sufficient for diagnostic purposes to conclude whether the system operates fault-free or not.

Second, fault identifiability may be lost by choosing wrong qualitative values. Consider the case where only the symbols low and normal are used to describe the continuous signal in the top most diagram of Fig. 3.1: neither parameter uncertainties nor the fault have an effect on the symbolic sequence (normal, low, normal). Hence, this description is not suitable for diagnostic purposes as the faulty behavior cannot be separated from the fault-free one.

Thus, the major engineering task in building qualitative models is to set up the conditions for the event generation which solve the conflict described above: They have to be chosen such that slight deviations in the continuous signals do not result in the generation of an event opposite to severe changes. Following this guideline, the information to set up a model is restricted to a minimum because there is no need to derive detailed models of the system in terms of differential equations.

3.2 Composite discrete-event systems

The system is considered in this contribution as the interconnection of a finite number of components C_i ($i \in \{1, \dots, \nu\}$) which are subject to individual faults $f_i \in \mathcal{N}_{f_i}$. The fault mode of the overall system is given by

$$\boldsymbol{f} = (f_1, \dots, f_\nu)^T \in \mathcal{N}_{\boldsymbol{f}} = \mathcal{N}_{f_1} \times \cdots \times \mathcal{N}_{f_\nu}. \tag{3.1}$$

Each component is subject to two different kinds of interactions (Fig. 3.2).

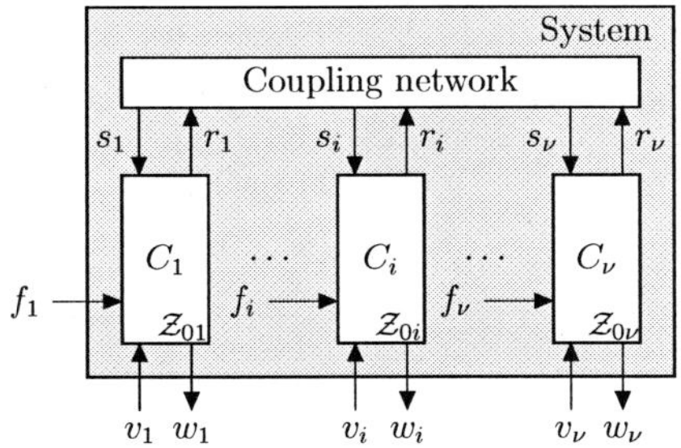

Figure 3.2: Composite discrete-event system subject to faults

First, the control input v_i and control output w_i are used to model the interaction of the component C_i with its environment. Correspondingly, this interaction is represented for the overall system by the signal vectors

$$\boldsymbol{v} = (v_1 \dots v_\nu)^T \in \mathcal{N}_{\boldsymbol{v}} = \{1, \dots, M\} = \mathcal{N}_{v_1} \times \dots \times \mathcal{N}_{v_\nu} \tag{3.2}$$

$$\boldsymbol{w} = (w_1 \dots w_\nu)^T \in \mathcal{N}_{\boldsymbol{w}} = \{1, \dots, R\} = \mathcal{N}_{w_1} \times \dots \times \mathcal{N}_{w_\nu}. \tag{3.3}$$

Second, two interconnection signals s_i and r_i are introduced for each component C_i to model the interaction with other components. These signals are assumed to be scalars to basically explain the modeling approach in Sections 3.3.1 to 3.3.7. The generalization to components with multiple inputs is given in Section 3.3.8. Accordingly, the signal vectors

$$\boldsymbol{s} = (s_1 \dots s_\nu)^T \in \mathcal{N}_{\boldsymbol{s}} = \{1, \dots, P\} = \mathcal{N}_{s_1} \times \dots \times \mathcal{N}_{s_\nu} \tag{3.4}$$
$$\boldsymbol{r} = (r_1 \dots r_\nu)^T \in \mathcal{N}_{\boldsymbol{r}} = \{1, \dots, T\} = \mathcal{N}_{r_1} \times \dots \times \mathcal{N}_{r_\nu} \tag{3.5}$$

are related to one another by the coupling network. By this, the coupling model does not contain any information about the dynamics of the system.

The behavior $\mathcal{B}_{C_i}$ of the component C_i is defined analogously to the behavior of the system given in eqn. (2.4) as the set of all control input, control output, interconnection input and interconnection output sequences of arbitrary length which can be generated by the component

$$\mathcal{B}_{C_i} \subseteq \mathcal{N}^*_{v_i} \times \mathcal{N}^*_{w_i} \times \mathcal{N}^*_{s_i} \times \mathcal{N}^*_{r_i}. \tag{3.6}$$

The actual measurements of the component are given as

$$B_i(k_e) = (V_i(0 \dots k_e), W_i(0 \dots k_e), S_i(0 \dots k_e), R_i(0 \dots k_e)). \tag{3.7}$$

According to eqn. (2.13), it is assumed that the model is identical to the component:

$$\mathcal{B}_{\mathcal{M}_i} = \mathcal{B}_{C_i}. \tag{3.8}$$

Hence, the component model is assumed to be complete and sound.

In the case of immeasurable interconnection signals, the outer behavior $\tilde{\mathcal{B}}_{C_i}$ of the component C_i is of interest. It is given by reducing eqn. (3.6) to

$$\tilde{\mathcal{B}}_{C_i} \subseteq \mathcal{N}^*_{v_i} \times \mathcal{N}^*_{w_i}. \tag{3.9}$$

Consequently, eqn. (3.7) reduces to

$$\tilde{B}_i(k_e) = (V_i(0 \dots k_e), W_i(0 \dots k_e)) \tag{3.10}$$

because only the control inputs and outputs are measurable.

3.3 Component models

I/O-automata are used throughout this work to model the components. The commonly used form can be found in [63, 70, 79]. It assumes that a state transition from z_i to a successor state z_i' with the according change of the output w_i is either due to a change in the input v_i, which forces this progress [117], or the free motion of the system which is known from continuous systems theory [71]. This set-up is called *event-driven* as opposed to a *time-driven* approach where a clock signal is used to frequently update the input, state and output. Changes in the state and the output occur *simultaneously* to the changes in the input. The integer number k counts the number of elapsed changes starting at a given reference point ($k = 0$). The fault f_i is used as an unknown input to influence this motion. The I/O-automata used here are *finite*, i.e. the set of states is finite. This type of I/O-automata is referred to as finite state automaton or finite state machine where automaton and state machine are synonyms [51]. The I/O-automata known from literature are extended in the following to cope with the interconnection input s_i and the interconnection output r_i.

3.3.1 Deterministic I/O-automata

The deterministic I/O-automaton $\mathcal{D}_i$ is used to model the behavior of each component C_i of a deterministic system subject to the fault f_i. The extension of I/O-automata to cope with the coupling signals is shown in Fig. 3.3.

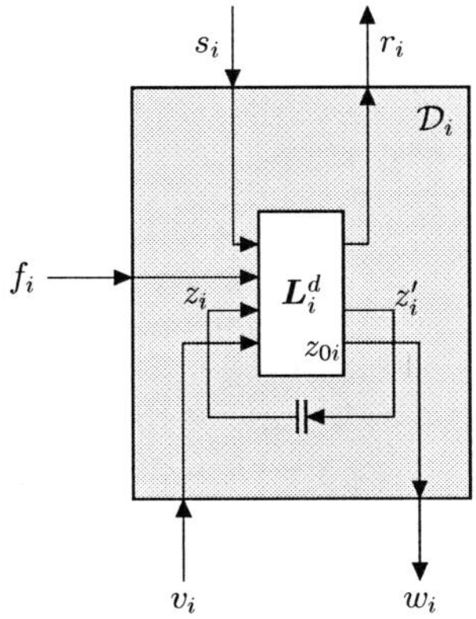

Figure 3.3: Structure of the deterministic I/O-automaton $\mathcal{D}_i$ subject to fault f_i

It results in describing the component by the model[1]

$$\mathcal{D}_i = (\mathcal{N}_{z_i}, \mathcal{N}_{v_i}, \mathcal{N}_{w_i}, \mathcal{N}_{s_i}, \mathcal{N}_{r_i}, \mathcal{N}_{f_i}, \boldsymbol{L}_i^d, z_{0i}) \tag{3.11}$$

[1]The superscript d is used for deterministic I/O-automata.

with the sets of

$$\mathcal{N}_{z_i} = \{1, \dots, N_i\} \quad \text{states} \tag{3.12}$$
$$\mathcal{N}_{v_i} = \{1, \dots, M_i\} \quad \text{control inputs} \tag{3.13}$$
$$\mathcal{N}_{w_i} = \{1, \dots, R_i\} \quad \text{control outputs} \tag{3.14}$$
$$\mathcal{N}_{s_i} = \{1, \dots, P_i\} \quad \text{interconnection inputs} \tag{3.15}$$
$$\mathcal{N}_{r_i} = \{1, \dots, T_i\} \quad \text{interconnection outputs and} \tag{3.16}$$
$$\mathcal{N}_{f_i} = \{0, \dots, S_i\} \quad \text{faults.} \tag{3.17}$$

These sets have a finite number of elements. The initial state is given by $z_{0i} \in \mathcal{N}_{z_i}$. Note that $\mathcal{N}_{z_i}$, $\mathcal{N}_{v_i}$, $\mathcal{N}_{w_i}$, $\mathcal{N}_{s_i}$, $\mathcal{N}_{r_i}$ and z_{0i} do not depend on the fault case f_i.

The dynamical properties of the component C_i are described by the behavioral relation $\boldsymbol{L}_i^d \subseteq \mathcal{N}_{z_i} \times \mathcal{N}_{w_i} \times \mathcal{N}_{r_i} \times \mathcal{N}_{z_i} \times \mathcal{N}_{v_i} \times \mathcal{N}_{s_i} \times \mathcal{N}_{f_i}$ of the I/O-automaton. $\boldsymbol{L}_i^d$ is defined by its characteristic function $\chi_{\boldsymbol{L}_i^d} : \mathcal{N}_{z_i} \times \mathcal{N}_{w_i} \times \mathcal{N}_{r_i} \times \mathcal{N}_{z_i} \times \mathcal{N}_{v_i} \times \mathcal{N}_{s_i} \times \mathcal{N}_{f_i} \to \{0, 1\}$ which has the value "1" for all tuples $(z_i(k+1), w_i(k), r_i(k), z_i(k), v_i(k), s_i(k), f_i)$ that specify a transition of the component subject to fault f_i from state $z_i(k)$ to the successor state $z_i(k+1)$ due to the control input $v_i(k)$ and interconnection input $s_i(k)$ while the control output $w_i(k)$ and the interconnection output $r_i(k)$ are produced. The value is equal to "0" in all other cases. Hence, the behavioral relation is given by

$$\begin{aligned} \boldsymbol{L}_i^d = \{ & (z_i(k+1), w_i(k), r_i(k), z_i(k), v_i(k), s_i(k), f_i) | \\ & \chi_{\boldsymbol{L}_i^d}(z_i(k+1), w_i(k), r_i(k), z_i(k), v_i(k), s_i(k), f_i) = 1\}. \end{aligned} \tag{3.18}$$

If the time k is not of interest, the abbreviations $\chi_{\boldsymbol{L}_i^d}(z_i', w_i, r_i, z_i, v_i, s_i, f_i)$ will be used. The symbol "||" in Fig. 3.3 represents a shift register that is used to store the successor state z_i' calculated from eqn. (3.18) for one time step.

The constraints on the diagnostic system described in Section 1.1 call for a model as small as possible, i.e. the model must only contain that part of the behavior of the system that is needed to fulfill the diagnostic task. Hence, it is suitable to deal with a characteristic function that is *partially* defined over its domain[2]. Control input values v_i and interconnection input values s_i for which a state transition is defined in the state z_i are called *activated*[3]. The associated sets of activated control input values $\mathcal{N}_{v_i}^{act}(z_i) \subseteq \mathcal{N}_{v_i}$ and activated interconnection input values $\mathcal{N}_{s_i}^{act}(z_i) \subseteq \mathcal{N}_{s_i}$ of the component C_i in the state z_i are defined as

$$\mathcal{N}_{v_i}^{act}(z_i) = \{v_i \in \mathcal{N}_{v_i} | \exists z_i', w_i, r_i, s_i, f_i : \chi_{\boldsymbol{L}_i^d}(z_i', w_i, r_i, z_i, v_i, s_i, f_i) = 1\} \tag{3.19}$$
$$\mathcal{N}_{s_i}^{act}(z_i) = \{s_i \in \mathcal{N}_{s_i} | \exists z_i', w_i, r_i, v_i, f_i : \chi_{\boldsymbol{L}_i^d}(z_i', w_i, r_i, z_i, v_i, s_i, f_i) = 1\}. \tag{3.20}$$

[2]Recall that the characteristic function is *totally* defined in discrete-event systems theory [27, 70].

[3]These values are called *active* or *feasible* in [27]. In there, the *active event function* Γ is used to define the set of active events in a state. But it is superfluous in the sense that Γ is derived from the state transition relation. Hence, it will be omitted here in the definition of the I/O-automaton.

Correspondingly, not all control output values w_i and interconnection output values r_i may be possible in each state. To retain a consistent notation, these values are referred to as activatable and the corresponding sets $\mathcal{N}_{w_i}^{act}(z_i) \subseteq \mathcal{N}_{w_i}$ and $\mathcal{N}_{r_i}^{act}(z_i) \subseteq \mathcal{N}_{r_i}$ are given as

$$\mathcal{N}_{w_i}^{act}(z_i) = \{w_i \in \mathcal{N}_{w_i} | \exists z_i', r_i, v_i, s_i, f_i : \chi_{\boldsymbol{L}_i^d}(z_i', w_i, r_i, z_i, v_i, s_i, f_i) = 1\} \tag{3.21}$$

$$\mathcal{N}_{r_i}^{act}(z_i) = \{r_i \in \mathcal{N}_{r_i} | \exists z_i', w_i, v_i, s_i, f_i : \chi_{\boldsymbol{L}_i^d}(z_i', w_i, r_i, z_i, v_i, s_i, f_i) = 1\}. \tag{3.22}$$

From the determinism follows that the behavioral relation is *well-posed* [4] as defined in Definition 9 on page 57:

$$\sum_{z_i'=1}^{N_i} \sum_{w_i=1}^{R_i} \sum_{r_i=1}^{T_i} \chi_{\boldsymbol{L}_i^d}(z_i', w_i, r_i, z_i, v_i, s_i, f_i) = 1 \tag{3.23}$$

$$\forall z_i \in \mathcal{N}_{z_i}, v_i \in \mathcal{N}_{v_i}^{act}(z_i), s_i \in \mathcal{N}_{s_i}^{act}(z_i), f_i \in \mathcal{N}_{f_i}.$$

The behavior of the deterministic I/O-automaton $\mathcal{D}_i$ is given for the fault f_i by the set $\mathcal{B}_{\mathcal{D}_i}(f_i)$ of tuples of control input, control output, interconnection input and interconnection output sequences that can be generated by the I/O-automaton:

$$\begin{aligned}\mathcal{B}_{\mathcal{D}_i}(f_i) := \{&(V_i(0\ldots k_e)/W_i(0\ldots k_e)/S_i(0\ldots k_e)/R_i(0\ldots k_e))| \\ &z_i(0) = z_{0i}, \exists Z_i(0\ldots k_e+1) : \\ &\boldsymbol{L}_i^d(z_i(k+1), w_i(k), r_i(k), z_i(k), v_i(k), s_i(k), f_i) = 1, 0 \leq k \leq k_e\}.\end{aligned} \tag{3.24}$$

The abbreviation

$$Z_i(0\ldots k_e+1) = (z_i(0), z_i(1), \ldots, z_i(k_e), z_i(k_e+1)) \tag{3.25}$$

is used in eqn. (3.24) to denote the state sequence. The behavior of the entire I/O-automaton includes for all faults all sequences of control inputs and outputs and interconnection inputs and outputs of arbitrary length for which the I/O-automaton can generate a state sequence given an initial state [20]. It is obtained by joining the behaviors for each fault f_i to one set:

$$\mathcal{B}_{\mathcal{D}_i} := \{\mathcal{B}_{\mathcal{D}_i}(f_i) | f_i \in \mathcal{N}_{f_i}\}. \tag{3.26}$$

In the case of immeasurable interconnection signals, the outer behavior of the deterministic I/O-automaton $\mathcal{D}_i$ is given for the fault f_i by the set $\tilde{\mathcal{B}}_{\mathcal{D}_i}(f_i)$ of tuples of control input and control output sequences that can be generated by the I/O-automaton:

$$\begin{aligned}\tilde{\mathcal{B}}_{\mathcal{D}_i}(f_i) := \{&(V_i(0\ldots k_e)/W_i(0\ldots k_e))|z_i(0) = z_{0i}, \exists Z_i(0\ldots k_e+1) : \\ &\sum_{r_i(k)=1}^{T_i} \sum_{s_i(k)=1}^{P_i} \boldsymbol{L}_i^d(z_i(k+1), w_i(k), r_i(k), z_i(k), v_i(k), s_i(k), f_i) = 1, 0 \leq k \leq k_e\}.\end{aligned} \tag{3.27}$$

The outer behavior of the entire I/O-automaton is given according to eqn. (3.26) by

$$\tilde{B}_{\mathcal{D}_i} := \{\tilde{\mathcal{B}}_{\mathcal{D}_i}(f_i) | f_i \in \mathcal{N}_{f_i}\}. \tag{3.28}$$

[4] In the case of totally defined state transition relations, eqn. (3.23) holds in every state $z_i \in \mathcal{N}_{z_i}$ and every fault $f_i \in \mathcal{N}_{f_i}$ for all $v_i \in \mathcal{N}_{v_i}$ and $s_i \in \mathcal{N}_{s_i}$ as $\mathcal{N}_{v_i}^{act}(z_i) = \mathcal{N}_{v_i}$ and $\mathcal{N}_{s_i}^{act}(z_i) = \mathcal{N}_{s_i}$.

3.3.2 Nondeterministic I/O-automata

The extension of deterministic to nondeterministic I/O-automata is useful in modeling due to the following reasons. First, it is possible to consider systems whose behavior cannot be predicted exactly. Second, the given information about the system may not be deep enough to model each effect of a change in the discrete signal with certainty. Third, even though the behavior of the system may be deterministic, representing it by a nondeterministic I/O-automaton is more space efficient than using the equivalent deterministic I/O-automaton [27, 70]. The nondeterministic I/O-automaton $\mathcal{N}_i$ with coupling signals is defined as[5]

$$\mathcal{N}_i = (\mathcal{N}_{z_i}, \mathcal{N}_{v_i}, \mathcal{N}_{w_i}, \mathcal{N}_{s_i}, \mathcal{N}_{r_i}, \mathcal{N}_{f_i}, \boldsymbol{L}_i^n, \mathcal{Z}_{0i}) \tag{3.29}$$

with the finite sets $\mathcal{N}_{z_i}$, $\mathcal{N}_{v_i}$, $\mathcal{N}_{w_i}$, $\mathcal{N}_{s_i}$, $\mathcal{N}_{r_i}$, $\mathcal{N}_{f_i}$ defined in (3.12) to (3.17). As the system is nondeterministic, the I/O-automaton may be initialized by the set of state $\mathcal{Z}_{0i}$ instead of a single state. Again, $\mathcal{N}_{z_i}$, $\mathcal{N}_{v_i}$, $\mathcal{N}_{w_i}$, $\mathcal{N}_{s_i}$, $\mathcal{N}_{r_i}$ and $\mathcal{Z}_{0i}$ do not depend on the fault case.

The dynamical properties of the component C_i are described by the behavioral relation $\boldsymbol{L}_i^n \subseteq \mathcal{N}_{z_i} \times \mathcal{N}_{w_i} \times \mathcal{N}_{r_i} \times \mathcal{N}_{z_i} \times \mathcal{N}_{v_i} \times \mathcal{N}_{s_i} \times \mathcal{N}_{f_i}$ of the I/O-automaton. $\boldsymbol{L}_i^n$ is defined by its characteristic function $\chi_{\boldsymbol{L}_i^n} : \mathcal{N}_{z_i} \times \mathcal{N}_{w_i} \times \mathcal{N}_{r_i} \times \mathcal{N}_{z_i} \times \mathcal{N}_{v_i} \times \mathcal{N}_{s_i} \times \mathcal{N}_{f_i} \rightarrow \{0, 1\}$ which has the value "1" for all *possible* transitions. The value is equal to "0" in all other cases. Hence, the behavioral relation is given by

$$\begin{aligned}\boldsymbol{L}_i^n = \{&(z_i(k+1), w_i(k), r_i(k), z_i(k), v_i(k), s_i(k), f_i) | \\ &\chi_{\boldsymbol{L}_i^n}(z_i(k+1), w_i(k), r_i(k), z_i(k), v_i(k), s_i(k), f_i) = 1\}.\end{aligned} \tag{3.30}$$

By this, the information structure depicted in Fig. 3.3 remains unchanged besides the use of $\boldsymbol{L}_i^n$ instead of $\boldsymbol{L}_i^d$ and $\mathcal{Z}_{0i}$ instead of z_{0i}. The abbreviations introduced in Section 3.3.1 can be applied to the nondeterministic case without any changes.

Again, the characteristic function is *partially* defined due to the consideration known from deterministic I/O-automata. The set of activated control input values $\mathcal{N}_{v_i}^{act}(z_i) \subseteq \mathcal{N}_{v_i}$ and activated interconnection input values $\mathcal{N}_{s_i}^{act}(z_i) \subseteq \mathcal{N}_{s_i}$ of the component C_i in the state z_i are defined as

$$\mathcal{N}_{v_i}^{act}(z_i) = \{v_i \in \mathcal{N}_{v_i} | \exists z_i', w_i, r_i, s_i, f_i : \chi_{\boldsymbol{L}_i^n}(z_i', w_i, r_i, z_i, v_i, s_i, f_i) = 1\} \tag{3.31}$$

$$\mathcal{N}_{s_i}^{act}(z_i) = \{s_i \in \mathcal{N}_{s_i} | \exists z_i', w_i, r_i, v_i, f_i : \chi_{\boldsymbol{L}_i^n}(z_i', w_i, r_i, z_i, v_i, s_i, f_i) = 1\}. \tag{3.32}$$

Correspondingly, the sets of activatable control output values $\mathcal{N}_{w_i}^{act}(z_i) \subseteq \mathcal{N}_{w_i}$ and activatable interconnection output values $\mathcal{N}_{r_i}^{act}(z_i) \subseteq \mathcal{N}_{r_i}$ are given as

$$\mathcal{N}_{w_i}^{act}(z_i) = \{w_i \in \mathcal{N}_{w_i} | \exists z_i', r_i, v_i, s_i, f_i : \chi_{\boldsymbol{L}_i^n}(z_i', w_i, r_i, z_i, v_i, s_i, f_i) = 1\} \tag{3.33}$$

$$\mathcal{N}_{r_i}^{act}(z_i) = \{r_i \in \mathcal{N}_{r_i} | \exists z_i', w_i, v_i, s_i, f_i : \chi_{\boldsymbol{L}_i^n}(z_i', w_i, r_i, z_i, v_i, s_i, f_i) = 1\}. \tag{3.34}$$

[5] The superscript n is used for nondeterministic I/O-automata.

The I/O-automaton is assumed to be *alive*

$$\bigvee_{z_i'=1}^{N_i} \bigvee_{w_i=1}^{R_i} \bigvee_{r_i=1}^{T_i} \chi_{\boldsymbol{L}_i^n}(z_i', w_i, r_i, z_i, v_i, s_i, f_i) = 1 \tag{3.35}$$
$$\forall z_i \in \mathcal{N}_{z_i}, v_i \in \mathcal{N}_{v_i}^{act}(z_i), s_i \in \mathcal{N}_{s_i}^{act}(z_i), f_i \in \mathcal{N}_{f_i}.$$

Hence, there exists at least one possible state transition for every activated control input, activated interconnection input and fault combination. The behavior and the outer behavior of the nondeterministic I/O-automaton $\mathcal{N}_i$ can be given according to eqns. (3.24), (3.26), (3.27) and (3.28).

3.3.3 I/O-automata without coupling signals

The I/O-automata defined in Sections 3.3.1 and 3.3.2 can appear without coupling signals. The resulting I/O-automata are only able to communicate with their environment but not with each other. Hence, the original form as defined in [63, 70] is obtained. From now on, *only nondeterministic I/O-automata are considered.* They are given without coupling signals as

$$\tilde{\mathcal{N}}_i = (\mathcal{N}_{z_i}, \mathcal{N}_{v_i}, \mathcal{N}_{w_i}, \mathcal{N}_{f_i}, \tilde{\boldsymbol{L}}_i^n, \mathcal{Z}_{0i}). \tag{3.36}$$

Their characteristic function $\chi_{\tilde{\boldsymbol{L}}_i^n} : \mathcal{N}_{z_i} \times \mathcal{N}_{w_i} \times \mathcal{N}_{z_i} \times \mathcal{N}_{v_i} \times \mathcal{N}_{f_i} \to \{0, 1\}$ has the value "1" for all *possible* transitions. The behavioral relation is defined as

$$\tilde{\boldsymbol{L}}_i^n = \{(z_i(k+1), w_i(k), z_i(k), v_i(k), f_i)| \\ \chi_{\tilde{\boldsymbol{L}}_i^n}(z_i(k+1), w_i(k), z_i(k), v_i(k), f_i) = 1\}. \tag{3.37}$$

It is *partially* defined due to the above mentioned reasons. Hence, the set of activated input values $\tilde{\mathcal{N}}_{v_i}^{act}(z_i) \subseteq \mathcal{N}_{v_i}$ and activatable control output values $\tilde{\mathcal{N}}_{v_i}(z_i) \subseteq \mathcal{N}_{w_i}$ are given in the state z_i as

$$\tilde{\mathcal{N}}_{v_i}^{act}(z_i) = \{v_i \in \mathcal{N}_{v_i} | \exists z_i', w_i, f_i : \chi_{\tilde{\boldsymbol{L}}_i^n}(z_i', w_i, z_i, v_i, f_i) = 1\}. \tag{3.38}$$
$$\tilde{\mathcal{N}}_{w_i}^{act}(z_i) = \{w_i \in \mathcal{N}_{w_i} | \exists z_i', v_i, f_i : \chi_{\tilde{\boldsymbol{L}}_i^n}(z_i', w_i, z_i, v_i, f_i) = 1\}. \tag{3.39}$$

The following properties of the characteristic functions hold according to eqn. (3.35):

$$\bigvee_{z_i'=1}^{N_i} \bigvee_{w_i=1}^{R_i} \chi_{\tilde{\boldsymbol{L}}_i^n}(z_i', w_i, z_i, v_i, f_i) = 1 \quad \forall z_i \in \mathcal{N}_{z_i}, v_i \in \tilde{\mathcal{N}}_{v_i}^{act}(z_i), f_i \in \mathcal{N}_{f_i}. \tag{3.40}$$

The behavior of the I/O-automaton without coupling signals is given by applying eqns. (3.24) and (3.26) accordingly.

3.3.4 Autonomous automata

The I/O-automata without coupling signals considered in the previous section can be further reduced to autonomous I/O-automata which have neither a control input, a control output nor a fault signal. They are not able to communicate with their environment. The nondeterministic autonomous I/O-automaton is given as

$$\widehat{\mathcal{N}}_i = (\mathcal{N}_{z_i}, \widehat{\boldsymbol{L}}_i^n, \mathcal{Z}_{0i}). \tag{3.41}$$

Its characteristic function $\chi_{\widehat{\boldsymbol{L}}_i^n} : \mathcal{N}_{z_i} \times \mathcal{N}_{z_i} \rightarrow \{0,1\}$ has the value "1" for all *possible* transitions and is used to define the behavioral relation as

$$\widehat{\boldsymbol{L}}_i^n = \{(z_i(k+1), z_i(k)) | \chi_{\widehat{\boldsymbol{L}}_i^n}(z_i(k+1), z_i(k)) = 1\}. \tag{3.42}$$

The following properties of the characteristic function hold according to eqn. (3.35):

$$\bigvee_{z_i'=1}^{N_i} \chi_{\widehat{\boldsymbol{L}}_i^n}(z_i', z_i) = 1 \quad \forall z_i \in \mathcal{N}_{z_i}. \tag{3.43}$$

The behavior of the autonomous I/O-automaton is given by applying eqns. (3.24) and (3.26) accordingly.

3.3.5 Interconnection relation, state transition relation and output relation

Three relations containing only partial information about the I/O-automaton can be extracted from the characteristic function of the behavioral relation. The **interconnection relation** $\boldsymbol{F}_i^d \subseteq \mathcal{N}_{r_i} \times \mathcal{N}_{z_i} \times \mathcal{N}_{v_i} \times \mathcal{N}_{s_i} \times \mathcal{N}_{f_i}$ contains information about the interconnection output r_i, the **state transition relation** $\boldsymbol{G}_i^d \subseteq \mathcal{N}_{z_i} \times \mathcal{N}_{z_i} \times \mathcal{N}_{v_i} \times \mathcal{N}_{s_i} \times \mathcal{N}_{f_i}$ about the successor state z_i' and the **output relation** $\boldsymbol{H}_i^d \subseteq \mathcal{N}_{w_i} \times \mathcal{N}_{z_i} \times \mathcal{N}_{v_i} \times \mathcal{N}_{s_i} \times \mathcal{N}_{f_i}$ about the control output w_i. Their characteristic functions $\chi_{\boldsymbol{F}_i^d} : \mathcal{N}_{r_i} \times \mathcal{N}_{z_i} \times \mathcal{N}_{v_i} \times \mathcal{N}_{s_i} \times \mathcal{N}_{f_i} \rightarrow \{0,1\}$, $\chi_{\boldsymbol{G}_i^d} : \mathcal{N}_{z_i} \times \mathcal{N}_{z_i} \times \mathcal{N}_{v_i} \times \mathcal{N}_{s_i} \times \mathcal{N}_{f_i} \rightarrow \{0,1\}$ and $\chi_{\boldsymbol{H}_i^d} : \mathcal{N}_{w_i} \times \mathcal{N}_{z_i} \times \mathcal{N}_{v_i} \times \mathcal{N}_{s_i} \times \mathcal{N}_{f_i} \rightarrow \{0,1\}$ can be derived from the characteristic function $\chi_{\boldsymbol{L}_i^d}$ in the case of deterministic I/O-automata by

$$\chi_{\boldsymbol{F}_i^d}(r_i, z_i, v_i, s_i, f_i) = \sum_{z_i'=1}^{N_i} \sum_{w_i=1}^{R_i} \chi_{\boldsymbol{L}_i^d}(z_i', w_i, r_i, z_i, v_i, s_i, f_i) \tag{3.44}$$

$$\chi_{\boldsymbol{G}_i^d}(z_i', z_i, v_i, s_i, f_i) = \sum_{w_i=1}^{R_i} \sum_{r_i=1}^{T_i} \chi_{\boldsymbol{L}_i^d}(z_i', w_i, r_i, z_i, v_i, s_i, f_i) \tag{3.45}$$

$$\chi_{\boldsymbol{H}_i^d}(w_i, z_i, v_i, s_i, f_i) = \sum_{z_i'=1}^{N_i} \sum_{r_i=1}^{T_i} \chi_{\boldsymbol{L}_i^d}(z_i', w_i, r_i, z_i, v_i, s_i, f_i). \tag{3.46}$$

In the nondeterministic case, the **interconnection relation** $\boldsymbol{F}_i^n \subseteq \mathcal{N}_{r_i} \times \mathcal{N}_{z_i} \times \mathcal{N}_{v_i} \times \mathcal{N}_{s_i} \times \mathcal{N}_{f_i}$, the **state transition relation** $\boldsymbol{G}_i^n \subseteq \mathcal{N}_{z_i} \times \mathcal{N}_{z_i} \times \mathcal{N}_{v_i} \times \mathcal{N}_{s_i} \times \mathcal{N}_{f_i}$ and the **output relation** $\boldsymbol{H}_i^n \subseteq \mathcal{N}_{w_i} \times \mathcal{N}_{z_i} \times \mathcal{N}_{v_i} \times \mathcal{N}_{s_i} \times \mathcal{N}_{f_i}$ are given by their characteristic functions $\chi_{\boldsymbol{F}_i^n} : \mathcal{N}_{r_i} \times \mathcal{N}_{z_i} \times \mathcal{N}_{v_i} \times \mathcal{N}_{s_i} \times \mathcal{N}_{f_i} \rightarrow \{0,1\}$, $\chi_{\boldsymbol{G}_i^n} : \mathcal{N}_{z_i} \times \mathcal{N}_{z_i} \times \mathcal{N}_{v_i} \times \mathcal{N}_{s_i} \times \mathcal{N}_{f_i} \rightarrow \{0,1\}$ and $\chi_{\boldsymbol{H}_i^n} : \mathcal{N}_{w_i} \times \mathcal{N}_{z_i} \times \mathcal{N}_{v_i} \times \mathcal{N}_{s_i} \times \mathcal{N}_{f_i} \rightarrow \{0,1\}$ which can be derived in the following way:

$$\chi_{\boldsymbol{F}_i^n}(r_i, z_i, v_i, s_i, f_i) = \bigvee_{z_i'=1}^{N_i} \bigvee_{w_i=1}^{R_i} \chi_{\boldsymbol{L}_i^n}(z_i', w_i, r_i, z_i, v_i, s_i, f_i) \tag{3.47}$$

$$\chi_{\boldsymbol{G}_i^n}(z_i', z_i, v_i, s_i, f_i) = \bigvee_{w_i=1}^{R_i} \bigvee_{r_i=1}^{T_i} \chi_{\boldsymbol{L}_i^n}(z_i', w_i, r_i, z_i, v_i, s_i, f_i) \tag{3.48}$$

$$\chi_{\boldsymbol{H}_i^n}(w_i, z_i, v_i, s_i, f_i) = \bigvee_{z_i'=1}^{N_i} \bigvee_{r_i=1}^{T_i} \chi_{\boldsymbol{L}_i^n}(z_i', w_i, r_i, z_i, v_i, s_i, f_i). \tag{3.49}$$

3.3.6 Automaton table and graph

Their are two equivalent ways to represent the behavioral relation of an I/O-automaton: The *automaton table* and the *automaton graph*. The first one basically lists all possible transitions of the I/O-automaton in a table. It is best suitable for the computer-based application. The columns of the table contain the variables of the characteristic function. The value of $\chi_{\boldsymbol{L}_i^n}$ is not explicitly given because only possible transitions are considered. The rows include the elements $(z_i', w_i, r_i, z_i, v_i, s_i, f_i)$ of $\boldsymbol{L}_i^n$ whereby the ordering is irrelevant. An example is given in Table 3.1.

Table 3.1: Exemplary automaton table

z_i'	w_i	r_i	z_i	v_i	s_i	f_i
2	1	2	1	2	2	0
1	1	2	1	2	2	0
1	2	1	2	1	1	0
1	1	2	2	2	2	0
1	1	2	1	2	2	1
2	1	1	1	1	2	1
2	1	2	2	2	2	1

The second representation is a graphical interpretation of the automaton table using a directed graph [37, 112]. The states are given as nodes and the state transitions as directed edges which are labeled with the tuple $(v_i, w_i, s_i, r_i, f_i)$ of control input, control output, interconnection input, interconnection output and fault (Fig. 3.4(a)). Figure 3.4(b) shows the automaton graph of the example from Table 3.1. Due to clarity, automaton graphs are sometimes given for each fault separately. This situation is depicted in Figs. 3.4(c) and 3.4(d) for the faults f_0 and f_1 given by the example.

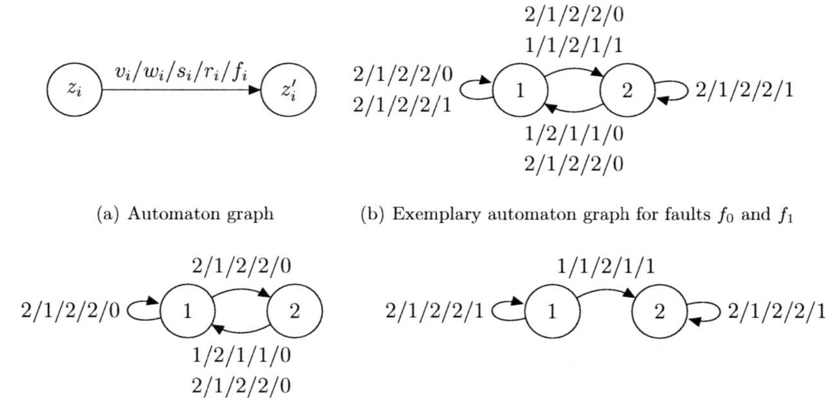

(a) Automaton graph

(b) Exemplary automaton graph for faults f_0 and f_1

(c) Exemplary automaton graph for fault f_0

(d) Exemplary automaton graph for fault f_1

Figure 3.4: Automaton graphs

3.3.7 Properties of the component models

Preliminaries. The *possibility function* $Poss(\sigma_{p1} = \sigma_1, \ldots, \sigma_{pn} = \sigma_n)$ introduced in [84] is used to assign a possibility represented by the numbers "0" and "1" to the simultaneous occurrence of the events σ_i $(i = 1, \ldots, n)$:

$$Poss(\sigma_1, \ldots, \sigma_n) \in \{0, 1\}. \tag{3.50}$$

It is formally given as $Poss(\sigma_{p1} = \sigma_1, \ldots, \sigma_{pn} = \sigma_n)$ which has the *logical value* "true" or "1", if and only if it is possible that all variables σ_{pi} assume the respective values σ_i *simultaneously*, else it is "false" or "0". The variables σ_{pi} are associated in this thesis with the signals of an I/O-automaton and the events σ_i with an element of the corresponding set. Thus, $Poss(\sigma_{pi} = \sigma_i) = 1$ denotes the *occurrence of this event*. The variables σ_{pi} will be omitted sometimes in $Poss(\sigma_{pi} = \sigma_i)$ for notational convenience. Instead, the abbreviation $Poss(\sigma_i)$ is used. Additionally, if the time k is not of interest, the abbreviations $Poss(z_i', w_i, r_i, z_i, v_i, s_i, f_i)$ will be used.

Markov Property. The Markov property is known from stochastic systems [23]. It indicates that the probability of a transition from a state to its successor state generating the control and interconnection output is *unambiguously* defined by the knowledge of the *current* state, control and interconnection input as well as the fault. In the case of nondeterministic systems, the Markov property indicates that the *set of successor states* is unambiguously defined. That is, the actual motion of the system is independent of the elapsed one:

$$\begin{aligned}&Poss(z_i(k+1), w_i(k), r_i(k), z_i(0), \ldots, z_i(k), v_i(0), \ldots, v_i(k),\\&\qquad s_i(0), \ldots, s_i(k), f_i(0), \ldots, f_i(k)) =\\&Poss(z_i(k+1), w_i(k), r_i(k), z_i(k), v_i(k), s_i(k), f_i(k)).\end{aligned} \tag{3.51}$$

This property holds for the deterministic and nondeterministic I/O-automaton defined in Sections 3.3.1 and 3.3.2[6]. In addition, the systems treated here are *homogeneous*, i.e. their behavioral relation is *time-invariant*:

$$\begin{aligned}&Poss(z_i(k+1) = z_i', w_i(k) = w_i, r_i(k) = r_i, z_i(k) = z_i, v_i(k) = v_i,\\&\qquad s_i(k) = s_i, f_i(k) = f_i) =\\&Poss(z_i(1) = z_i', w_i(0) = w_i, r_i(0) = r_i, z_i(0) = z_i, v_i(0) = v_i,\\&\qquad s_i(0) = s_i, f_i(0) = f_i).\end{aligned} \tag{3.52}$$

Special classes of I/O-automata. The *Mealy* automaton defined in [80] and the *Moore* automaton defined in [82] are of interest in this thesis. The first one assumes a direct dependence of the output upon the actual input and the current state, whereas the output in the second case depends only on the current state [70]. These concepts can be applied to the I/O-automata defined in Sections 3.3.1 and 3.3.2 with regard to the analysis presented in Section 4.1 w.r.t the interconnection relation $\boldsymbol{F}_i^n$ as follows.

Definition 5. *(Mealy property w.r.t. the interconnection signals) A nondeterministic I/O-automaton $\mathcal{N}_i$ is said to have the* Mealy property w.r.t. its interconnection signals, *if its interconnection relation $\boldsymbol{F}_i^n$ depends on $z_i(k)$, $v_i(k)$, $s_i(k)$ and $f_i(k)$:*

$$r_i(k) \in \boldsymbol{F}_i^n(r_i(k), z_i(k), v_i(k), s_i(k), f_i(k)). \tag{3.53}$$

Hence, the interconnection output $r_i(k)$ is directly influenced by the control input $v_i(k)$, the interconnection input $s_i(k)$ and the fault $f_i(k)$. It is related to the corresponding state transition $z_i(k) \rightarrow z_i(k+1)$. ◊

Definition 6. *(Moore property w.r.t. the interconnection signals) A nondeterministic I/O-automaton $\mathcal{N}_i$ is said to have the* Moore property w.r.t. its interconnection signals, *if its interconnection relation $\boldsymbol{F}_i^n$ does not depend on $s_i(k)$:*

$$r_i(k) \in \boldsymbol{F}_i^n(r_i(k), z_i(k), v_i(k), f_i(k)). \tag{3.54}$$

Hence, the interconnection output $r_i(k)$ is directly influenced only by the control input $v_i(k)$ and the fault $f_i(k)$. ◊

[6] Recall that the Markov property is, from a system theoretical point of view, defined for stochastic processes. They are referred to as *Markov chains*. The extension to the deterministic and nondeterministic case is shown e.g. in [70].

3.3.8 Extension to I/O-automata with multiple signals

The I/O-automaton structure shown in Fig. 3.3 is restrictive in the sense that the behavior of the component may depend on *one* external source and may influence *one* external sink. The same considerations hold for the internal couplings: The component reacts to the signal of *one* component and affects *another* one. This information structure can be generalized to cope with multiple interactions as shown in Fig. 3.5 by introducing the signal vectors

$$\begin{aligned}
\boldsymbol{v}_i &= (v_i^1, \dots, v_j^1, \dots, v_i^{\mu_i})^T \in \mathcal{N}_{\boldsymbol{v}_i} = \{1, \dots, M_i\} && (3.55)\\
&= \mathcal{N}_{v_i^1} \times \dots \times \mathcal{N}_{v_i^j} \times \dots \times \mathcal{N}_{v_i^{\mu_i}}\\
&= \{1, \dots, M_i^1\} \times \dots \times \{1, \dots, M_i^j\} \times \dots \times \{1, \dots, M_i^{\mu_i}\},\\
\boldsymbol{w}_i &= (w_i^1, \dots, w_i^j, \dots, w_i^{\rho_i})^T \in \mathcal{N}_{\boldsymbol{w}_i} = \{1, \dots, R_i\} && (3.56)\\
&= \mathcal{N}_{w_i^1} \times \dots \times \mathcal{N}_{w_i^j} \times \dots \times \mathcal{N}_{w_i^{\rho_i}}\\
&= \{1, \dots, R_i^1\} \times \dots \times \{1, \dots, R_i^j\} \times \dots \times \{1, \dots, R_i^{\rho_i}\},\\
\boldsymbol{s}_i &= (s_i^1, \dots, s_i^j, \dots, s_i^{\pi_i})^T \in \mathcal{N}_{\boldsymbol{s}_i} = \{1, \dots, P_i\} && (3.57)\\
&= \mathcal{N}_{s_i^1} \times \dots \times \mathcal{N}_{s_i^j} \times \dots \times \mathcal{N}_{s_i^{\pi_i}}\\
&= \{1, \dots, P_i^1\} \times \dots \times \{1, \dots, P_i^j\} \times \dots \times \{1, \dots, P_i^{\pi_i}\},\\
\boldsymbol{r}_i &= (r_i^1, \dots, r_i^j, \dots, r_i^{\tau_i})^T \in \mathcal{N}_{\boldsymbol{r}_i} = \{1, \dots, T_i\} && (3.58)\\
&= \mathcal{N}_{r_i^1} \times \dots \times \mathcal{N}_{r_i^j} \times \dots \times \mathcal{N}_{r_i^{\tau_i}}\\
&= \{1, \dots, T_i^1\} \times \dots \times \{1, \dots, T_i^j\} \times \dots \times \{1, \dots, T_i^{\tau_i}\}
\end{aligned}$$

for the control and the interconnection signals, where μ_i denotes the number of control inputs, ρ_i the number of control outputs, π_i the number of interconnection inputs and τ_i the number of interconnection outputs of the ith component.

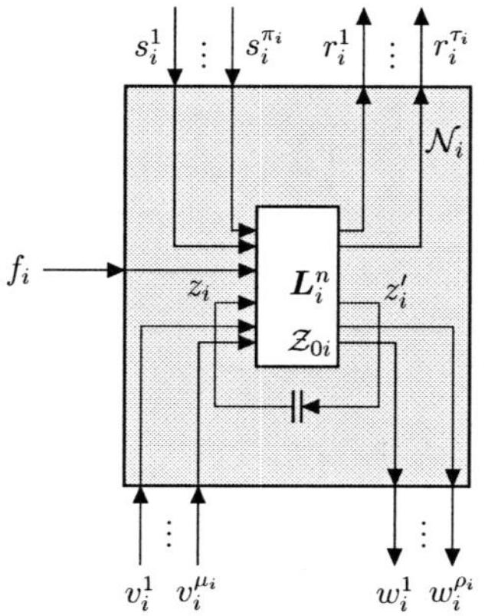

Figure 3.5: Generalized structure of the I/O-automaton with multiple inputs and outputs

The resulting nondeterministic I/O-automaton is given by

$$\mathcal{N}_i = (\mathcal{N}_{z_i}, \mathcal{N}_{\boldsymbol{v}_i}, \mathcal{N}_{\boldsymbol{w}_i}, \mathcal{N}_{\boldsymbol{s}_i}, \mathcal{N}_{\boldsymbol{r}_i}, \mathcal{N}_{f_i}, \boldsymbol{L}_i^n, \mathcal{Z}_{0i}) \tag{3.59}$$

with the behavioral relation

$$\boldsymbol{L}_i^n \subseteq \mathcal{N}_{z_i} \times \mathcal{N}_{\boldsymbol{w}_i} \times \mathcal{N}_{\boldsymbol{r}_i} \times \mathcal{N}_{z_i} \times \mathcal{N}_{\boldsymbol{v}_i} \times \mathcal{N}_{\boldsymbol{s}_i} \times \mathcal{N}_{f_i}. \tag{3.60}$$

3.4 Coupling model

The interaction block with the coupling model $\boldsymbol{K}$ is introduced to explicitly cover the internal structure of the system which is given by the coupling network in Fig. 3.2. It describes solely the links between the interconnection signals and does not contain any dynamical properties of the system. The basic concepts are studied in Section 3.4.1 assuming scalar interconnection signals. The generalization of the coupling model to multiple internal signals is given in Section 3.4.2.

3.4.1 I/O-automata with scalar interconnection signals

The coupling model is represented by the matrix $\boldsymbol{K}$ that links the interconnection outputs given by eqn. (3.4) with the interconnection inputs given by (3.5):

$$\boldsymbol{s} = \boldsymbol{K} \cdot \boldsymbol{r}. \tag{3.61}$$

This matrix is square because the system consists of ν components with exactly one interconnection input and one interconnection output. It has one element K_{ij} in each row that has the value "1" to model the coupling $s_i = r_j$. All other elements have the value "0". Hence, the relation

$$s_i = \boldsymbol{k}_i^T \cdot \boldsymbol{r} \quad \forall\, i \in \{1, \ldots, \nu\} \tag{3.62}$$

holds with $\boldsymbol{k}_i^T$ being the ith row of the matrix $\boldsymbol{K}$. By this, every component may affect several other components via its interconnection output because one output signal may be connected to several input signals in discrete-event system theory. The assumption to use scalar interaction signals restricts the interconnection in the opposite direction since the influence of several output signals on one input signal is not defined. Hence, there may be components that have no effect on any other component. The resulting problem of unconnected coupling signals will be discussed in Section 3.7.

Example 2 *Two different interconnection structures*

Two different examples for the interconnection of three components are given in Fig. 3.6.

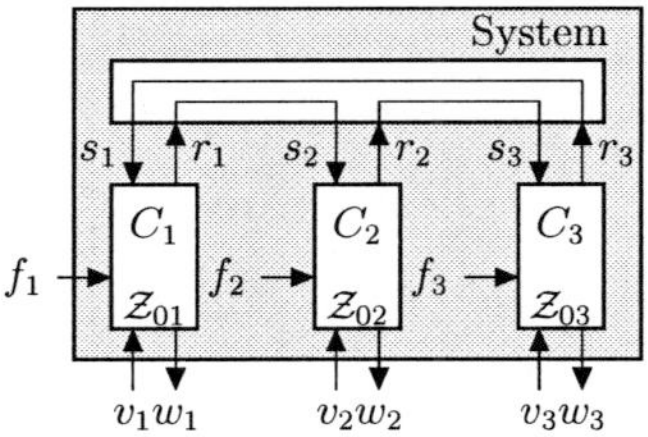

(a) Interconnection 1

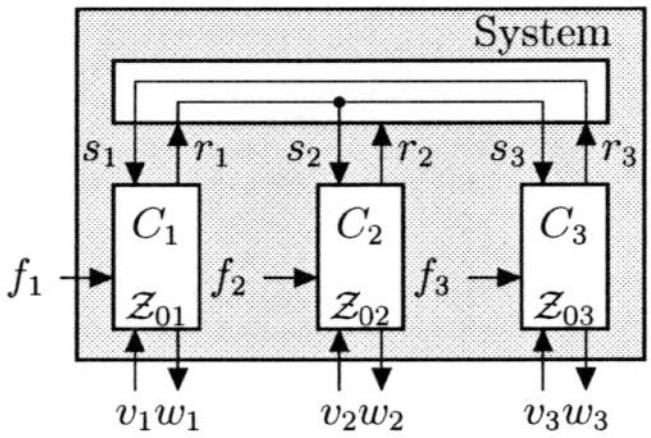

(b) Interconnection 2

Figure 3.6: Two different interconnection structures

The first one assumes an interconnection where all components affect each other (Fig. 3.6(a)). The corresponding coupling model is given by

$$\boldsymbol{K} = \begin{pmatrix} 0 & 0 & 1 \\ 1 & 0 & 0 \\ 0 & 1 & 0 \end{pmatrix}. \tag{3.63}$$

The second one considers the case where component C_1 influences components C_2 and *C_3. Thus, either component C_2 or C_3 cannot have an effect on another component due to the limitations described above. As seen from Fig. 3.6(b), component C_2 is the one on this example. This interaction is modeled by*

$$\boldsymbol{K} = \begin{pmatrix} 0 & 0 & 1 \\ 1 & 0 & 0 \\ 1 & 0 & 0 \end{pmatrix}. \tag{3.64}$$

3.4.2 Generalized I/O-automata

The interaction constraints resulting from scalar coupling signals can be overcome by modeling the component using the generalized I/O-automaton given by eqn. (3.59). Hence, the overall interconnection input and output are given by the signal vectors

$$\boldsymbol{s}^T = \left(\boldsymbol{s}_1^T, \ldots, \boldsymbol{s}_i^T, \ldots, \boldsymbol{s}_\nu^T\right)^T = \left(s_1^1, \ldots, s_1^{\pi_1}, \ldots, s_i^1, \ldots, s_i^{\pi_i}, \ldots, s_\nu^1, \ldots, s_\nu^{\pi_\nu}\right)^T, \tag{3.65}$$

$$\boldsymbol{r}^T = \left(\boldsymbol{r}_1^T, \ldots, \boldsymbol{r}_i^T, \ldots, \boldsymbol{r}_\nu^T\right)^T = \left(r_1^1, \ldots, r_1^{\tau_1}, \ldots, r_i^1, \ldots, r_i^{\tau_i}, \ldots, r_\nu^1, \ldots, r_\nu^{\tau_\nu}\right)^T. \tag{3.66}$$

These generalized interconnection signals can be used in conjunction with eqn. (3.61) to model the interaction between the components. The resulting coupling model $\boldsymbol{K}$ is a $\left(\sum_{i=1}^{\nu} \pi_i\right) \times \left(\sum_{i=1}^{\nu} \tau_i\right)$-matrix. It has *exactly one element* $K_{\tilde{i}\tilde{j}} = 1$ *in each row* with $\tilde{i} \in \left\{1, \ldots, \sum_{i=1}^{\nu} \pi_i\right\}$ and $\tilde{j} \in \left\{1, \ldots, \sum_{i=1}^{\nu} \tau_i\right\}$. The other elements are equal to zero such that

$$s_{\tilde{i}} = \boldsymbol{k}_{\tilde{i}}^T \cdot \boldsymbol{r} \quad \forall\, \tilde{i} \in \left\{1, \ldots, \sum_{i=1}^{\nu} \pi_i\right\} \tag{3.67}$$

holds for all interconnection inputs with $\boldsymbol{k}_{\tilde{i}}^T$ being the $\tilde{i}^{th}$ row of $\boldsymbol{K}$.

Example 3 *A general interconnection structure*

One example for a generalized coupling structure is given by the production facility from [70] considered in Section 8.2 in detail. This system consists of five components. The control unit C_1 contains the sorting task. It influences the robot C_2 to place two different products, each arriving on its own conveyor C_3 and C_4, in a defined order on the leaving conveyor C_5. The resulting interconnection is depicted in Fig. 3.7.

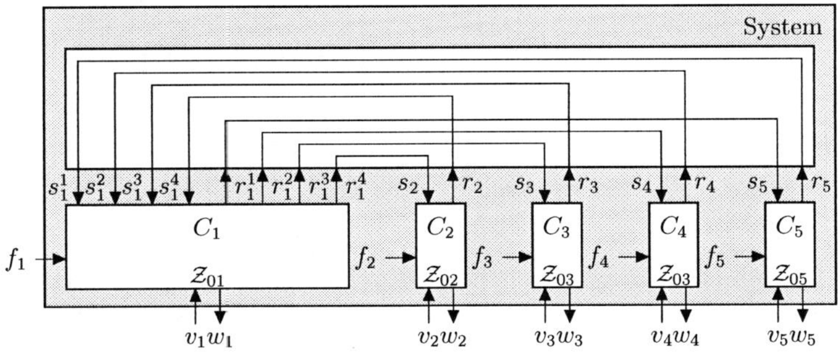

Figure 3.7: A general interconnection structure

The coupling relation between these components is given by

$$\begin{pmatrix} s_1^1 \\ s_1^2 \\ s_1^3 \\ s_1^4 \\ s_2 \\ s_3 \\ s_4 \\ s_5 \end{pmatrix} = \begin{pmatrix} 0 & 0 & 0 & 0 & 0 & 0 & 0 & 1 \\ 0 & 0 & 0 & 0 & 0 & 0 & 1 & 0 \\ 0 & 0 & 0 & 0 & 0 & 1 & 0 & 0 \\ 0 & 0 & 0 & 0 & 1 & 0 & 0 & 0 \\ 0 & 0 & 0 & 1 & 0 & 0 & 0 & 0 \\ 0 & 0 & 1 & 0 & 0 & 0 & 0 & 0 \\ 0 & 1 & 0 & 0 & 0 & 0 & 0 & 0 \\ 1 & 0 & 0 & 0 & 0 & 0 & 0 & 0 \end{pmatrix} \cdot \begin{pmatrix} r_1^1 \\ r_1^2 \\ r_1^3 \\ r_1^4 \\ r_2 \\ r_3 \\ r_4 \\ r_5 \end{pmatrix}. \tag{3.68}$$

3.5 Overall system model

The structured model of the overall system is introduced in Section 3.5.1 as networks of I/O-automata. It is given in the most general case: The internal coupling signals may depend directly and instantaneously through a signal path within the network on themselves. This so-called feedback connection may result in algebraic loops. Solutions to this problem are given in Section 4.1 in detail. It will be assumed in the remainder of this thesis that all networks of I/O-automata are free of conflicts. The composition of the I/O-automata network to an equivalent I/O-automaton without coupling signals is examined in Section 3.5.2 for the deterministic and the nondeterministic case.

3.5.1 Description of the network

The model of the overall system is given in the deterministic case by the deterministic asynchronous network of I/O-automata $\mathcal{DAN}$

$$\mathcal{DAN} = (\{\mathcal{D}_1, \ldots, \mathcal{D}_\nu\}, \mathcal{N}_{\boldsymbol{z}}, \mathcal{N}_{\boldsymbol{v}}, \mathcal{N}_{\boldsymbol{w}}, \mathcal{N}_{\boldsymbol{s}}, \mathcal{N}_{\boldsymbol{r}}, \mathcal{N}_{\boldsymbol{f}}, \boldsymbol{K}, \boldsymbol{z}_0) \tag{3.69}$$

with the network state

$$\boldsymbol{z} = (z_1 \ldots z_\nu)^T \in \mathcal{N}_{\boldsymbol{z}} = \mathcal{N}_{z_1} \times \cdots \times \mathcal{N}_{z_\nu}, \tag{3.70}$$

the initial state

$$\boldsymbol{z}_0 = (z_{0_1}, \ldots, z_{0_\nu})^T \tag{3.71}$$

and the fault given by eqn. (3.1). The control input $\boldsymbol{v} \in \mathcal{N}_{\boldsymbol{v}}$ and the control output $\boldsymbol{w} \in \mathcal{N}_{\boldsymbol{w}}$ are used to describe the interaction of the system with its environment. The interconnection input $\boldsymbol{s} \in \mathcal{N}_{\boldsymbol{s}}$ and the interconnection output $\boldsymbol{r} \in \mathcal{N}_{\boldsymbol{r}}$ are not seen from outside the system. The coupling of the components is explicitly described by the coupling model $\boldsymbol{K}$ as introduced in Section 3.4.2. Each component is given as the deterministic I/O-automaton $\mathcal{D}_i$ with the form (3.11).

In the nondeterministic case, the nondeterministic asynchronous network of I/O-automata $\mathcal{NAN}$ is given by

$$\mathcal{NAN} = (\{\mathcal{N}_1, \ldots, \mathcal{N}_\nu\}, \mathcal{N}_{\boldsymbol{z}}, \mathcal{N}_{\boldsymbol{v}}, \mathcal{N}_{\boldsymbol{w}}, \mathcal{N}_{\boldsymbol{s}}, \mathcal{N}_{\boldsymbol{r}}, \mathcal{N}_{\boldsymbol{f}}, \boldsymbol{K}, \mathcal{Z}_0) \tag{3.72}$$

with the network state introduced in eqn. (3.70) and the fault given by eqn. (3.1). As each component is described by the nondeterministic I/O-automaton $\mathcal{N}_i$ defined in eqn. (3.29), there is a set of initial states

$$\mathcal{Z}_0 = \mathcal{Z}_{01} \times \ldots \times \mathcal{Z}_{0\nu} \tag{3.73}$$

instead of a single initial state as in the deterministic case described above. The resulting structure is shown in Fig. 3.8.

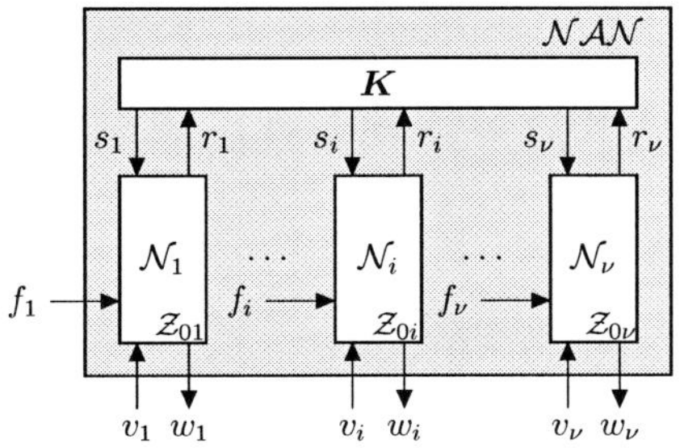

Figure 3.8: Network of I/O-automata in block diagram form subject to faults

3.5.2 Composition of I/O-automata networks

The aggregation of the network of I/O-automata to an equivalent single I/O-automaton without coupling signals which has the same input/output behavior and set of state sequences as the network is shown in this section. This operation is called *composition* [63, 70]. The composition rules given here are an extension of the results in [72, 84, 113] to I/O-automata with coupling signals where Definition 7 has been taken from [84] and has been extended here. The construction procedure presented in this section extends [72] to networks of I/O-automata in the most general form allowing for loops within the network. The components are assumed to have scalar signals within this section for simplicity of notation.

Definition 7. *(Equivalence) Two discrete-event systems S_1 and S_2 are said to be* equivalent, *if their I/O-behaviors given by eqn.* (3.26) *are identical:*

$$\mathcal{B}_{\mathcal{D}_1} = \mathcal{B}_{\mathcal{D}_2}. \tag{3.74}$$

◊

Deterministic behavior. The deterministic asynchronous network of I/O-automata $\mathcal{DAN}$ given by eqn. (3.69) has an equivalent representation as the single I/O-automaton $\tilde{\mathcal{D}} = (\mathcal{N}_{\boldsymbol{z}}, \mathcal{N}_{\boldsymbol{v}}, \mathcal{N}_{\boldsymbol{w}}, \mathcal{N}_{\boldsymbol{f}}, \tilde{\boldsymbol{L}}^d, \boldsymbol{z}_0)$ with the behavioral relation $\tilde{\boldsymbol{L}}^d \subseteq \mathcal{N}_{\boldsymbol{z}} \times \mathcal{N}_{\boldsymbol{w}} \times \mathcal{N}_{\boldsymbol{z}} \times \mathcal{N}_{\boldsymbol{v}} \times \mathcal{N}_{\boldsymbol{f}}$. It is an extension of the component model without coupling signals to multiple signals. A state transition

$$\begin{pmatrix} z_1 \\ z_2 \\ \vdots \\ z_\nu \end{pmatrix} \xrightarrow{\begin{pmatrix} v_1 \\ v_2 \\ \vdots \\ v_\nu \end{pmatrix} / \begin{pmatrix} w_1 \\ w_2 \\ \vdots \\ w_\nu \end{pmatrix} / \begin{pmatrix} f_1 \\ f_2 \\ \vdots \\ f_\nu \end{pmatrix}} \begin{pmatrix} z_1' \\ z_2' \\ \vdots \\ z_\nu' \end{pmatrix} \quad \Leftrightarrow \quad \chi_{\tilde{\boldsymbol{L}}^d}(\boldsymbol{z}', \boldsymbol{w}, \boldsymbol{z}, \boldsymbol{v}, \boldsymbol{f}) = 1 \tag{3.75}$$

occurs in the overall system, if all components perform a state transition

$$\forall i \in \{1,\ldots,\nu\}: \; z_i \xrightarrow{v_i/w_i/s_i/r_i/f_i} z_i' \quad \Leftrightarrow \quad \prod_{i=1}^{\nu} \chi_{\boldsymbol{L}_i^d}(z_i', w_i, r_i, z_i, v_i, s_i, f_i) = 1. \tag{3.76}$$

The interconnection input of the ith component is coupled according to eqn. (3.62) with an interconnection output of another component such that the right part of eqn. (3.76) can be rewritten as

$$\prod_{i=1}^{\nu} \chi_{\boldsymbol{L}_i^d}(z_i', w_i, r_i, z_i, v_i, \boldsymbol{k}_i^T \cdot \boldsymbol{r}, f_i) = 1. \tag{3.77}$$

The characteristic function $\chi_{\tilde{\boldsymbol{L}}^d}$ does not depend on the interconnection outputs as they cannot be seen from outside the system. Eliminating them from eqn. (3.77) is done by evaluating this equation for all possible interconnection outputs $\boldsymbol{r} \in \mathcal{N}_{\boldsymbol{r}}$. Thus, the transition of the overall system given by eqn. (3.75) can be obtained by the component models by applying the composition rule

$$\boxed{\chi_{\tilde{\boldsymbol{L}}^d}(\boldsymbol{z}', \boldsymbol{w}, \boldsymbol{z}, \boldsymbol{v}, \boldsymbol{f}) = \sum_{\boldsymbol{r}=1}^{T} \prod_{i=1}^{\nu} \chi_{\boldsymbol{L}_i^d}(z_i', \boldsymbol{w}_i, r_i, z_i, \boldsymbol{v}_i, \boldsymbol{k}_i^T \cdot \boldsymbol{r}, f_i)} \tag{3.78}$$

to the characteristic functions of the component models for all activated control inputs $\boldsymbol{v} \in \tilde{\mathcal{N}}_{\boldsymbol{v}}^{act}(\boldsymbol{z})$ and faults $\boldsymbol{f} \in \mathcal{N}_{\boldsymbol{f}}$ in each state $\boldsymbol{z} \in \mathcal{N}_{\boldsymbol{z}}$.

Nondeterministic behavior. The nondeterministic asynchronous network of I/O-automata $\mathcal{NAN}$ given by eqn. (3.72) can be represented analogously to the deterministic case by the single I/O-automaton $\tilde{\mathcal{N}} = (\mathcal{N}_{\boldsymbol{z}}, \mathcal{N}_{\boldsymbol{v}}, \mathcal{N}_{\boldsymbol{w}}, \mathcal{N}_{\boldsymbol{f}}, \tilde{\boldsymbol{L}}^n, \mathcal{Z}_0)$. The characteristic function $\chi_{\tilde{\boldsymbol{L}}^n}$ of the behavioral relation $\tilde{\boldsymbol{L}}^n$ can be obtained by the component models by applying the composition rule

$$\boxed{\chi_{\tilde{\boldsymbol{L}}^n}(\boldsymbol{z}', \boldsymbol{w}, \boldsymbol{z}, \boldsymbol{v}, \boldsymbol{f}) = \bigvee_{\boldsymbol{r}=1}^{T} \bigwedge_{i=1}^{\nu} \chi_{\boldsymbol{L}_i^n}(z_i', \boldsymbol{w}_i, r_i, z_i, \boldsymbol{v}_i, \boldsymbol{k}_i^T \cdot \boldsymbol{r}, f_i).} \tag{3.79}$$

to the characteristic functions of the component models for all activated control inputs $\boldsymbol{v} \in \tilde{\mathcal{N}}_{\boldsymbol{v}}^{act}(\boldsymbol{z})$ and faults $\boldsymbol{f} \in \mathcal{N}_{\boldsymbol{f}}$ in each state $\boldsymbol{z} \in \mathcal{N}_{\boldsymbol{z}}$. An example for the composition of a system consisting of two components is given on page 60.

3.6 The notion of asynchrony

The components of a network of I/O-automata move *synchronously* due to the assumption that a control or interconnection input value results in a simultaneous state transition and the corresponding generation of a control and interconnection output value. The interconnection output value influences directly the connected component. The assumption that all signal changes occur instantaneously, i.e. without consuming any time, results in the synchronous motion of all connected components. Synchronous motion is not always given in coupled mechatronical systems because there may be operating points that require the interaction of some but not necessarily all components (Section 1.1). Hence, a way to deal with this *asynchronous* motion has to be included in the new modeling approach.

In modeling interconnected discrete-event systems by standard automata, asynchrony means that one component may carry out a state transition *independently* of the other components, if a *private event* of the component triggers the system [27, 70]. The remaining components do not perform any state transition. To cover asynchrony, several composition operators like the *parallel composition* have been defined.

This situation occurs for I/O-automata that are coupled in the *side-by-side-composition* introduced in [63]. There, the *stuttering symbol* called `absent` is used to allow for the reaction of a single component in the composition. The reacting component gets a non-vanishing input symbol due to which it performs a state transition and generates an output symbol. The stuttering symbol is given to the other component to remain in its current state and not to output a new symbol. As a result, one component performs an independent state transition. Recall that the stuttering symbol is also called the *empty symbol* and denoted by ε.

The formalism of the empty symbol ε is extended in this thesis to cope with asynchrony in the new modeling formalism. It is introduced as an additional element in the sets (3.55) to (3.58):

$$\mathcal{N}_{v_i^j} = \mathcal{N}_{v_i^j} \cup \{\varepsilon\}, \tag{3.80}$$

$$\mathcal{N}_{w_i^j} = \mathcal{N}_{w_i^j} \cup \{\varepsilon\}, \tag{3.81}$$

$$\mathcal{N}_{s_i^j} = \mathcal{N}_{s_i^j} \cup \{\varepsilon\}, \tag{3.82}$$

$$\mathcal{N}_{r_i^j} = \mathcal{N}_{r_i^j} \cup \{\varepsilon\}. \tag{3.83}$$

Extending the generalized component models like this may result asynchronous effects:

- **Asynchronous state transition.** A component C_i performs an asynchronous state transition $z_i \rightarrow z_i'$ and generates the control output $\boldsymbol{w}_i$ due to the non-vanishing control input $\boldsymbol{v}_i$, if it gets the vector of empty interconnection input symbols $\boldsymbol{s}_i = \boldsymbol{\varepsilon}$ and generates the vector of empty interconnection output symbols $\boldsymbol{r}_i = \boldsymbol{\varepsilon}$ such that the relation

$$\exists\, z_i', \boldsymbol{w}_i : \boldsymbol{L}_i^n(z_i', \boldsymbol{w}_i, \boldsymbol{\varepsilon}, z_i, \boldsymbol{v}_i, \boldsymbol{\varepsilon}, f_i) = 1 \tag{3.84}$$

  holds for all $f_i \in \mathcal{N}_{f_i}$. Hence, the motion of the system is originated externally and there is no influence among the components.

- **No state transition.** A component C_i does neither change its state, i.e. $z_i' = z_i$, nor generate a control output symbol due to the vector of empty control input symbols

$\boldsymbol{v}_i = \boldsymbol{\varepsilon}$, if the condition

$$\boldsymbol{L}_i^n(z_i, \boldsymbol{\varepsilon}, \boldsymbol{r}_i, z_i, \boldsymbol{\varepsilon}, \boldsymbol{s}_i, f_i) = 1 \tag{3.85}$$

holds on the behavioral relation for all $z_i \in \mathcal{N}_{z_i}$, $\boldsymbol{s}_i \in \mathcal{N}_{\boldsymbol{s}_i}$, $\boldsymbol{r}_i \in \mathcal{N}_{\boldsymbol{r}_i}$ and $f_i \in \mathcal{N}_{f_i}$.

Correspondingly, the following synchronous situations may occur:

- **State transition caused by another component.** The motion of a component C_i from state z_i to z_i' while generating the control output $\boldsymbol{w}_i$ and the interconnection output $\boldsymbol{r}_i$ is forced by another component C_j, if C_i gets the vector of empty control input symbols $\boldsymbol{v}_i = \boldsymbol{\varepsilon}$ and the non-vanishing interconnection input $\boldsymbol{s}_i$. Thus, the following condition holds for some $z_i \in \mathcal{N}_{z_i}$, $\boldsymbol{s}_i \in \mathcal{N}_{\boldsymbol{s}_i}$ and $f_i \in \mathcal{N}_{f_i}$:

  $$\exists\ z_i', \boldsymbol{w}_i, \boldsymbol{r}_i : \boldsymbol{L}_i^n(z_i', \boldsymbol{w}_i, \boldsymbol{r}_i, z_i, \boldsymbol{\varepsilon}, \boldsymbol{s}_i, f_i) = 1. \tag{3.86}$$

  The component performs a state transition that is not seen from the outside, if it produces the empty control output symbol.

- **Blocked state transition.** A transition of the component C_i can be blocked by the interconnection input $\boldsymbol{s}_i$, i.e. by another component C_j, even though it gets a non-vanishing control input $\boldsymbol{v}_i$ such that the behavioral relation is given as

  $$\boldsymbol{L}_i^n(z_i, \boldsymbol{\varepsilon}, \boldsymbol{\varepsilon}, z_i, \boldsymbol{v}_i, \boldsymbol{s}_i, f_i) = 1 \tag{3.87}$$

  for all $z_i \in \mathcal{N}_{z_i}$, $\boldsymbol{v}_i \in \mathcal{N}_{\boldsymbol{v}_i}$, $\boldsymbol{s}_i \in \mathcal{N}_{\boldsymbol{s}_i}$ and $f_i \in \mathcal{N}_{f_i}$.

Asynchrony of one component given by eqn. (3.84) can be extended to several components, if there is an interaction between these components due to non-vanishing interconnection signals given by eqn. (3.86). The participating components may get in addition non-vanishing control inputs, i.e. their motion may be caused by their interconnection input or additionally by their control input. In this case, the participating components perform a *synchronous* state transition while the non-participating components remain in their current states. Hence, the interacting components move *asynchronously* with respect to the remaining components. If all components are part of the interaction, the whole system performs a synchronous state transition. The properties (3.85) and (3.87) may be satisfied for some instead of all states of the components or inputs such that the corresponding phenomena do not hold always but in this specific situations. These different kinds of state transitions are illustrated in Example 4 on page 60 in conjunction with aspects of composition.

3.7 Unconnected signals of I/O-automata

The definitions of the deterministic I/O-automaton given by eqn. (3.11) and of the nondeterministic I/O-automaton given by eqn. (3.29) allows for four different kinds of interaction: The control input and output as well as the interconnection input and output. These signals may be scalars or vectors (Section 3.3.8). Thus, multiple interactions have already been defined. This section is concerned with the opposite question on how to deal with signals that are not needed in the current configuration.

The motivation of this task arises from the wish to create a library of component models such that the description of a given system can be easily obtained by taking these component models and linking them via an interconnection block to the overall system model. The challenge in this context is to wind up with a model of a certain type of component that is applicable to a possibly large variety of this type of component. Hence, there might have been signals taken into account via modeling that are irrelevant in the actual setting.

An example may be a crossing point for pipes. It may connect two, three, four or possibly more pipes with another such that the number of interconnection inputs and outputs corresponds to the number of coupled pipes. As the current application may require the coupling of three pipes, some interconnection signals are unconnected. This is indicated in the coupling model by setting the ith row vector $\boldsymbol{k}_i^T$ equal to $\boldsymbol{0}^T$. Thus, the component model must be able to handle a *permanent* "0" at the corresponding interconnection input. In addition, the corresponding interconnection output is also unconnected such that the ith column vector $\boldsymbol{k}_i$ is equal to $\boldsymbol{0}$. This case is uncritical because it does not influence the behavior of the component. This discussion can be copied directly to the control signals. Hence, the task is to extend the new modeling formalism such that it is able to handle a permanent "0" at a control or interconnection input. Thus, the definitions (3.80) and (3.82) have to be extended by

$$\mathcal{N}_{\boldsymbol{v}_i} = \mathcal{N}_{\boldsymbol{v}_i} \cup \{0\} \tag{3.88}$$

$$\mathcal{N}_{\boldsymbol{s}_i} = \mathcal{N}_{\boldsymbol{s}_i} \cup \{0\}. \tag{3.89}$$

The set of control outputs $\mathcal{N}_{\boldsymbol{w}_i}$ and interconnection outputs $\mathcal{N}_{\boldsymbol{r}_i}$ do not have to be extended as done in [11] because unconnecting these signals results in an open prot which has no influence on anything. The empty symbol cannot be used in this context because it indicates the asynchronous motion of a component which has to be distinguished from unconnected signals. The following situations may result from this extension of the component models:

- **Unconnected interconnection input.** The behavioral relation of the component must be defined for

$$\boldsymbol{L}_i^n(z_i', \boldsymbol{w}_i, \boldsymbol{r}, z_i, \boldsymbol{v}_i, \boldsymbol{0}, f_i) = 1 \tag{3.90}$$

during the modeling process. It may be used e.g. to either define a series or a feedback connection with another component. The interconnection input is unconnected in the case of a series connection. Hence, the resulting influence of this component onto the other one has to be modeled by defining the interconnection outputs correspondingly. If the interconnection input is connected, the behavior of the component depends on the interconnection input and must be given accordingly.

- **Unconnected control input.** The behavioral relation of the component has to be defined for

$$\boldsymbol{L}_i^n(z_i', \boldsymbol{w}_i, \boldsymbol{r}, z_i, \boldsymbol{0}, \boldsymbol{s}_i, f_i) = 1 \tag{3.91}$$

during the modeling process. The same considerations as for the above case hold for the control input because there may be situations where the component is influenced by its environment ($\boldsymbol{v}_i \neq \boldsymbol{0}$) or not ($\boldsymbol{v}_i = \boldsymbol{0}$).

- **Unconnected control and interconnection input.** The behavioral relation of the component must be defined for

$$\boldsymbol{L}_i^n(z_i', \boldsymbol{w}_i, \boldsymbol{r}, z_i, \boldsymbol{0}, \boldsymbol{0}, f_i) = 1 \tag{3.92}$$

during the modeling process. Thus, only the free motion of the component has to be modeled.

3.8 Comparison of networks of I/O-automata with coupled standard automata

Standard automata are widely used in discrete-event systems theory. They distinguish from the model introduced in the previous section by defining interactions among the components in terms of events rather than interconnection input and output symbols. These events are also used to represent state transitions [27, 63, 133]. For component-oriented modeling, several composition operators have been proposed in literature [131, 132]. The main idea is to distinguish private events that may occur in a single component, from common events that have to occur in all components. The *synchronous product* is used to describe the synchronization of the behavior of two components for all common events, whereas the *parallel composition* permits autonomous state changes of single components for their private events. For systems with higher priority events, the *prioritized synchronous composition* can be used [17, 50].

The comparison of networks of I/O-automata with coupled standard automata is presented in the following for the parallel composition of two deterministic components C_1 and C_2 without a fault signal $\boldsymbol{f} \in \mathcal{N}_{\boldsymbol{f}}$. Therefore, the modeling of discrete-event systems by coupled deterministic standard automata is introduced in Section 3.8.1. The interpretation of deterministic networks of I/O-automata is given in Section 3.8.2. The equivalence of both modeling formalisms is shown in Section 3.8.3 by an example. The formal proof of equivalence including the extension to systems with more than two components and nondeterministic behavior can be found in [73]. This section closes with a discussion of the results.

3.8.1 Modeling of discrete-event systems by standard automata

The component C_i $(i = 1, 2)$ is modeled by the deterministic standard automaton $\mathcal{S}_i$, which is defined by the 4-tuple

$$\mathcal{S}_i = (\mathcal{N}_{z_i}, \Sigma_i, \boldsymbol{\delta}_i, z_{0i}) \tag{3.93}$$

where $\mathcal{N}_{z_i}$ is the set of states, $z_{0i} \in \mathcal{N}_{z_i}$ the initial state and Σ_i the set of events. The state transition function $\boldsymbol{\delta}_i : \mathcal{N}_{z_i} \times \Sigma_i \rightarrow \mathcal{N}_{z_i}$ specifies in each state z_i for all events σ_i the successor state z_i'

$$z_i' = \boldsymbol{\delta}_i(z_i, \sigma_i). \tag{3.94}$$

For the composition of the two components, the *parallel composition rule* is used. It yields the overall system model

$$\mathcal{S}_{\|} = (\mathcal{N}_{z_{\|}}, \Sigma_{\|}, \boldsymbol{\delta}_{\|}, \boldsymbol{z}_{0\|}) \tag{3.95}$$

with the state $\boldsymbol{z}_{\|} = (z_1, z_2)^T \in \mathcal{N}_{z_{\|}} = \mathcal{N}_{z_1} \times \mathcal{N}_{z_2}$, the initial state $\boldsymbol{z}_{0\|} = (z_{01}, z_{02})^T$ and the set of events $\Sigma_{\|} = \Sigma_1 \cup \Sigma_2$. The introduction of the sets of common and private events

$$\Sigma^{com} = \Sigma_1 \cap \Sigma_2, \; i = 1, 2 \tag{3.96}$$

$$\Sigma_i^{priv} = \Sigma_i \backslash \Sigma^{com}, \; i = 1, 2 \tag{3.97}$$

is needed to define the state transition function $\boldsymbol{\delta}_{\|}$ of the overall system

$$\boldsymbol{\delta}_{\|}\left(\begin{pmatrix} z_1 \\ z_2 \end{pmatrix}, \sigma\right) = \begin{cases} \begin{pmatrix} \boldsymbol{\delta}_1(z_1, \sigma) \\ \boldsymbol{\delta}_2(z_2, \sigma) \end{pmatrix}, & \text{if } \boldsymbol{\delta}_1(z_1, \sigma)! \wedge \; \boldsymbol{\delta}_2(z_2, \sigma)! \quad (3.98a) \\ \begin{pmatrix} \boldsymbol{\delta}_1(z_1, \sigma) \\ z_2 \end{pmatrix}, & \text{if } \sigma \in \Sigma_1^{priv} \wedge \; \boldsymbol{\delta}_1(z_1, \sigma)! \quad (3.98b) \\ \begin{pmatrix} z_1 \\ \boldsymbol{\delta}_2(z_2, \sigma) \end{pmatrix}, & \text{if } \sigma \in \Sigma_2^{priv} \wedge \; \boldsymbol{\delta}_2(z_2, \sigma)! \quad (3.98c) \\ \text{undefined}, & \text{otherwise} \quad (3.98d) \end{cases}$$

that yields the successor state $\boldsymbol{z}_{\|}'$. The notation $\boldsymbol{\delta}_i(z_i, \sigma)!$ means that the state transition function $\boldsymbol{\delta}_i$ is defined for the arguments z_i and σ.

3.8.2 Interpretation of asynchronous networks of I/O-automata as networks of standard automata

This section shows that the network of I/O-automata can be used to describe a discrete-event system modeled as the parallel composition of standard automata. As other composition rules can likewise be modeled by networks of I/O-automata, the new model introduced in this thesis is at least applicable for the same class of systems as known modeling methods that use standard automata.

The events occurring in standard automata are interpreted as inputs to I/O-automata (like in the case where standard automata are used as acceptors of a language) to relate both modeling classes to one another. It is assumed that the same input $v \in \mathcal{N}_{v_{\|}} = \mathcal{N}_{v_1} \cup \mathcal{N}_{v_2}$ is assigned to both components, which results in the special structure of the network of I/O-automata shown in Fig. 3.9.

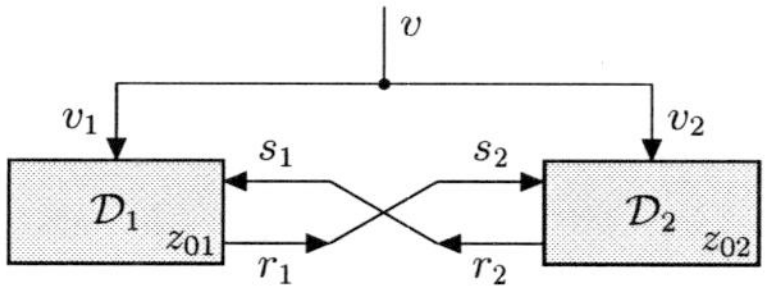

Figure 3.9: Interpretation of a network of I/O-automata as coupled standard automata

The interconnection signals s_i and r_i are used to enable or disable the movement of the I/O-automaton $\mathcal{D}_i$ by $\mathcal{D}_j$ and vice versa to re-build the parallel composition of two standard automata. Consequently, the interconnection inputs and outputs need to have two elements: $\mathcal{N}_{s_i} = \mathcal{N}_{r_i} = \{0, 1\}$ $(i = 1, 2)$. If $s_i = 1$ holds, the I/O-automaton $\mathcal{D}_i$ is allowed to move, which means that for this argument the state transition function $\boldsymbol{G}_i$ is defined. Otherwise ($s_i = 0$), the movement of $\mathcal{D}_i$ is prohibited by I/O-automaton $\mathcal{D}_j$ with the consequence, that $\boldsymbol{G}_i$ is not defined for this argument. Analogously, $r_i = 0$ means that $\mathcal{D}_i$ sends a prohibition to $\mathcal{D}_j$ and allows for the movement of $\mathcal{D}_j$ by $r_i = 1$.

The parallel composition rule given by eqn. (3.98) is realized by an appropriate choice of the interconnection function $\boldsymbol{F}_i$ of the I/O-automata as follows:

Definition 8. *(Interconnection function for the parallel composition) The interconnection function $\boldsymbol{F}_i^{pac}$ is defined by*

$$\boldsymbol{F}_i^{pac}(z_i, v_i) = r_i = \begin{cases} 1, & \text{if } \boldsymbol{G}_i(z_i, v_i, 1)! \;\vee\; v_i \notin \mathcal{N}_{v_i} \quad (3.99a) \\ 0, & \textit{otherwise.} \quad (3.99b) \end{cases}$$

◊

Interpretation of eqn. (3.99). The I/O-automaton $\mathcal{D}_i$ allows the movement of I/O-automaton $\mathcal{D}_j$, if its state transition function $\boldsymbol{G}_i$ is defined or if $v_i \notin \mathcal{N}_{v_i}$. The negation of this statement yields the condition for permitting the movement. $\boldsymbol{G}_i$ is not defined and the input symbol is not known to the I/O-automaton $\mathcal{D}_i$, i.e. it is a private input symbol of the I/O-automaton $\mathcal{D}_j$.

As the network of I/O I/O-automata considered here has no control output, the I/O-automaton $\mathcal{D}_i$ is defined without an output function by the 7-tuple

$$\mathcal{D}_i = (\mathcal{N}_{z_i}, \mathcal{N}_{v_i}, \mathcal{N}_{s_i}, \mathcal{N}_{r_i}, \boldsymbol{F}_i^{pac}, \boldsymbol{G}_i, z_{0i}) \qquad (3.100)$$

with the state transition function $\boldsymbol{G}_i : \mathcal{N}_{z_i} \times \mathcal{N}_{v_i} \times \mathcal{N}_{s_i} \rightarrow \mathcal{N}_{z_i}$ which calculates in each state z_i the successor state z_i' for all control inputs v_i and interconnection inputs s_i

$$z_i' = \boldsymbol{G}_i(z_i, v_i, s_i). \tag{3.101}$$

Its structure is depicted in Fig. 3.10.

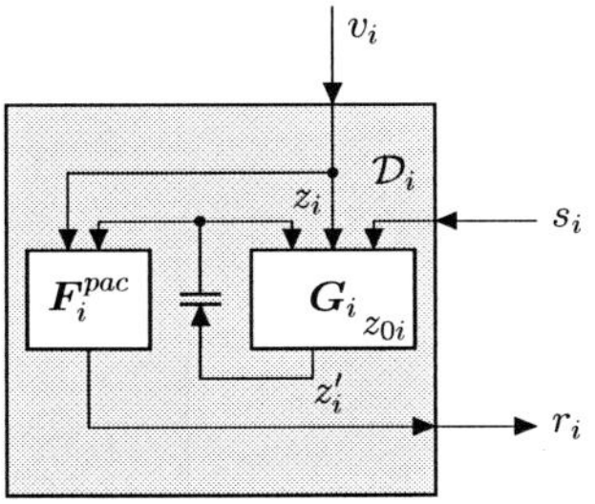

Figure 3.10: Structure of the reduced I/O-automaton

The I/O-automaton $\mathcal{D}_{\|}$ representing the parallel composition of $\mathcal{D}_1$ und $\mathcal{D}_2$ is defined by the 4-tuple

$$\mathcal{D}_{\|} = (\mathcal{N}_{z_{\|}}, \mathcal{N}_{v_{\|}}, \boldsymbol{G}_{\|}, \boldsymbol{z}_{0\|}) \tag{3.102}$$

with the set of states $\mathcal{N}_{z_{\|}} = \mathcal{N}_{z_1} \times \mathcal{N}_{z_2}$ and the initial state $\boldsymbol{z}_{0\|} = (z_{01}, z_{02})^T$. The state transition function of the network of I/O-automata

$$\boldsymbol{G}_{\|} : \mathcal{N}_{z_{\|}} \times \mathcal{N}_{\boldsymbol{v}} \rightarrow \mathcal{N}_{z_{\|}} \tag{3.103}$$

is obtained by the following procedure. First, the value of the interconnection outputs r_1 and r_2 are calculated by eqn. (3.99) in each state $\boldsymbol{z}_{\|} \in \mathcal{N}_{z_{\|}}$ for each input $v \in \mathcal{N}_{v_{\|}}$ and linked to the interconnection inputs as shown in Fig. 3.10. Next, the successor state $\boldsymbol{z}_{\|}' = (z_1', z_2')^T$ is calculated using eqn. (3.101). Hence, the state transition $\boldsymbol{z}_{\|}' = \boldsymbol{G}_{\|}(\boldsymbol{z}_{\|}, v)$ is possible for the overall system.

3.8.3 Example

The comparison of the modeling formalisms is illustrated by the example shown in Fig. 3.11. If the automata graphs are interpreted for the standard automata, a directed edge from state z_i to z_i' labeled with σ_i is used for all possible state transitions $z_i' = \boldsymbol{\delta}_i(z_i, \sigma_i)$. For the interpretation as I/O-automata, the edge is labeled with the pair (v_i/s_i) for all possible state transitions $z_i' = \boldsymbol{G}_i(z_i, v_i, s_i)$. In addition, it contains in addition the value of the interconnection input.

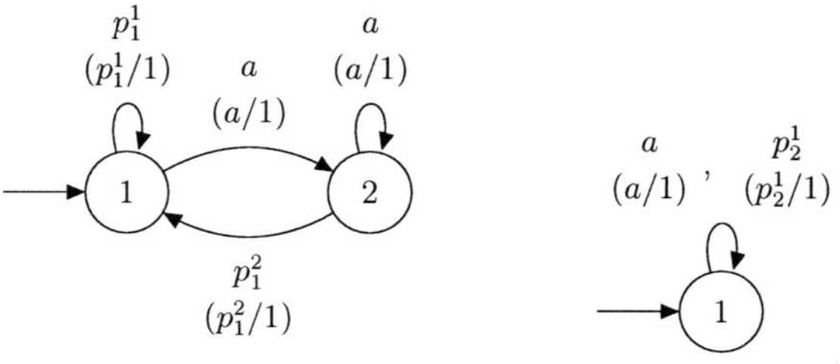

(a) Automaton $\mathcal{S}_1$ (resp. $\mathcal{N}_1$) (b) Automaton $\mathcal{S}_2$ (resp. $\mathcal{N}_2$)

Figure 3.11: Automaton graph of the standard automata $\mathcal{S}_i$ (and the I/O-automata $\mathcal{N}_i$)

The sets of events of the components are chosen to be

$$\Sigma_1 = \{a, p_1^1, p_1^2\} = \mathcal{N}_{v_1} \qquad \Sigma_2 = \{a, p_2^1\} = \mathcal{N}_{v_2}. \tag{3.104}$$

Note that common events have the same name. Private events are denoted by p_i^x where the index i corresponds with the component C_i and the superscript $x = 1, 2, \ldots$ is used consecutively to enumerate the events of the component:

$$\Sigma^{com} = \{a\} = \mathcal{N}_{\boldsymbol{v}}^{com} \tag{3.105}$$

$$\Sigma_1^{priv} = \{p_1^1, p_1^2\} = \mathcal{N}_{v_1}^{priv} \qquad \Sigma_2^{priv} = \{p_2^1\} = \mathcal{N}_{v_2}^{priv}. \tag{3.106}$$

The set of events of the overall system is given by

$$\Sigma_{\|} = \{a, p_1^1, p_1^2, p_2^1\} = \mathcal{N}_{v_{\|}} \tag{3.107}$$

and the set of states $\mathcal{N}_{z_{\|}}$ by

$$\mathcal{N}_{z_{\|}} = \left\{(1,1)^T, (2,1)^T\right\} \tag{3.108}$$

with the initial state $\boldsymbol{z}_{0\|} = (z_{01}, z_{02})^T = (1,1)^T$.

The standard automaton $\mathcal{S}_{\|}$ depicted in Fig. 3.12 is obtained by the application of the parallel composition rule given by eqn. (3.98) to $\mathcal{S}_1$ and $\mathcal{S}_2$.

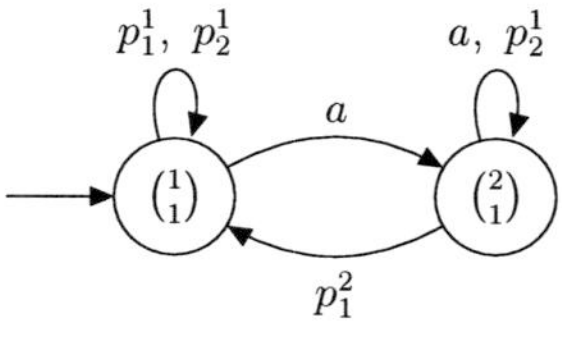

Figure 3.12: Automaton graph of the composed standard automaton $\mathcal{S}_{\|}$

The construction procedure to obtain the I/O-automaton $\mathcal{D}_{\|}$ is shown step-by-step in Tab. 3.2. Comparing this results with the standard automaton depicted in Fig. 3.12 shows the equality of the two models.

Table 3.2: Construction of the composed I/O-automaton $\mathcal{N}_\|$

$\boldsymbol{z}_\| = \binom{z_1}{z_2}$	v	$r_1 = s_2$	$r_2 = s_1$	$\boldsymbol{G}_1(\cdot)!$?	$\boldsymbol{G}_2(\cdot)!$?	$\boldsymbol{z}'_\| = \binom{z'_1}{z'_2}$	Entry in $\boldsymbol{G}_\|$
$(1,1)^T$	a	1	1	y	y	$(2,1)^T$	$\binom{2}{1} = \boldsymbol{G}_\| \left(\binom{1}{1}, a\right)$
$(1,1)^T$	p_1^1	1	1	y	n	$(1,1)^T$	$\binom{1}{1} = \boldsymbol{G}_\| \left(\binom{1}{1}, p_1^1\right)$
$(1,1)^T$	p_1^2	0	1	n	n	-	-
$(1,1)^T$	p_2^1	1	1	n	y	$(1,1)^T$	$\binom{1}{1} = \boldsymbol{G}_\| \left(\binom{1}{1}, p_1^2\right)$
$(2,1)^T$	a	1	1	y	y	$(2,1)^T$	$\binom{2}{1} = \boldsymbol{G}_\| \left(\binom{2}{1}, a\right)$
$(2,1)^T$	p_1^1	0	1	n	n	-	-
$(2,1)^T$	p_1^2	1	1	y	n	$(1,1)^T$	$\binom{1}{1} = \boldsymbol{G}_\| \left(\binom{2}{1}, p_2^1\right)$
$(2,1)^T$	p_2^1	1	1	n	y	$(2,1)^T$	$\binom{2}{1} = \boldsymbol{G}_\| \left(\binom{2}{1}, p_1^2\right)$

3.8.4 Discussion

Section 3.8.3 has shown that it is possible to define a *special class* of networks of I/O-automata introduced in Sections 3.2 to 3.5 to re-build the parallel composition rule known from modeling of discrete-event systems by coupled standard automata. This fact points to three major advantages of the modeling formalism introduced in this thesis in comparison to the standard automata modeling approach:

1. The new modeling formalism can be proven to be powerful enough to cover all composition operators known from standard automata modeling (for a survey on composition operators see e.g. [131]).

2. The cause-and-effect-chains of the system under consideration are explicitly modeled by networks of I/O-automata. For standard automata modeling, the cause-and-effect-chains of the system are implicitly covered in the choice of the events.

3. The way of modeling standard automata is a *top-down* approach. The most important modeling step and the major disadvantage with respect to I/O-automata modeling is the requirement that the events of the whole system have to be chosen and categorized from an overall system point of view. This procedure violates the principles of component-oriented modeling demanded in this thesis as opposed to the modeling formalism introduced in Sections 3.2 to 3.5 which can be used *bottom-up* by defining the component models completely separately.

Chapter 4

Properties and simulation of asynchronous networks of I/O-automata

Different aspects of the analysis of the new modeling formalism are presented in this chapter. Solutions to the feedback problem arising from algebraic loops within the network are given in Section 4.1. In Section 4.2, two different types of autonomy are introduced. Two algorithms for the simulation of the behavior of nondeterministic asynchronous networks of I/O-automata are given in Section 4.3. This chapter closes with complexity considerations in Section 4.4.

4.1 Well-posedness

The network of I/O-automata defined in Chapter 3 consists of several interconnected components. State transitions are assumed to be forced by either control or interconnection input values or the free motion of the system. The resulting changes in the state and the generation of output values occur simultaneously. Hence, the motion of the system is synchronized and may, therefore, become ill-defined, if one or more I/O-automata form algebraic loops such that the state transitions depend directly and instantaneously upon each other through direct feedback paths within the network. This so-called *direct feedback problem* may lead to conflicts [63]. It is investigated in Section 4.1.1 on the results presented in [70, 84] for synchronous networks of I/O-automata. Solutions have been developed in [11] for the deterministic case and published in [6] on an algorithmic base. The closed form of this algorithm is presented and formally proven in Section 4.1.2 and extended to the nondeterministic case.

4.1.1 Feedback in interconnected systems

Direct feedback occurs in coupled systems when a direct dependency of a signal y on itself is given:

$$y = f(y). \tag{4.1}$$

This problem requires to find the *fixed-points* of the relation (4.1). In general, there may be no, one or several fixed-points but only a unique solution $\bar{y} = f(\bar{y})$ is called *well-posed* [63].

The "algebraic loop" given by eqn. (4.1) arises from replacing $s_i = \boldsymbol{k}_i^T \cdot \boldsymbol{r}$ in the interconnection relations (3.44). The overall interconnection relation

$$\boldsymbol{r} = \boldsymbol{F}^d(\boldsymbol{r}, \boldsymbol{z}, \boldsymbol{v}, \boldsymbol{K} \cdot \boldsymbol{r}) \tag{4.2}$$

shows the direct dependency of the interconnection output on itself. A conflict occurs whenever the interconnection relation forces the interconnection output on the left-hand side of eqn. (4.2) to assume a value different from that one on the right-hand side.

Algebraic loops do not exist in reality. They arise from abstracting information during modeling to obtain a simplified description of the system. The resulting component models may be suitable, if they are considered independently. But conflicts may occur, if these component models are considered in conjunction. The used level of abstraction may result in the simultaneous appearance of an input and an output value in the model such that relation (4.2) holds even though there is a temporal ordering between their generation in the real system. It is obvious that this problem is restricted to synchronized systems and does not occur, if these changes may be evaluated at different points of time [62].

In addition, the direct feedback problem arises only in components which allow for a *direct feed-through* of the interconnection input s_i to the interconnection output r_i. Direct feed-through can be avoided, if a component has the *Moore property w.r.t. the interconnection signals* given by Definition 6. Hence, the following lemma from [63] can be stated without proof for asynchronous networks of I/O-automata.

Lemma 1. *A loop in an network of I/O-automata is well-posed, if it contains at least one I/O-automaton that has the Moore property w.r.t. the interconnection signals.* ◇

The general case of I/O-automata allowing for a direct feed-through is investigated in the next section.

4.1.2 Solution to the feedback problem for asynchronous networks of I/O-automata

Direct feed-through occurs in networks of I/O-automata where the component models have the *Mealy property w.r.t. the interconnection signals* given by Definition 5. Conditions are investigated in the following to ensure the well-posedness of deterministic and nondeterministic asynchronous networks of I/O-automata. Scalar signals are used within this section for simplicity of notation.

Deterministic behavior. The formal composition of a deterministic network of I/O-automata to an equivalent I/O-automaton without coupling signals is given by eqn. (3.78).

Definition 9. *(Deterministic composition) The deterministic asynchronous network of I/O-automata $\mathcal{DAN}$ (3.69) is called* well-posed, *if and only if the following conditions hold:*

1. *The relation*

$$\mathcal{N}_{r_i} \subseteq \mathcal{N}_{s_j} \tag{4.3}$$

holds for all interconnection inputs and interconnection outputs that are related by the coupling model $\boldsymbol{K}$ to one another.

2. *There exists for all activated control inputs $\boldsymbol{v} \in \tilde{\mathcal{N}}_{\boldsymbol{v}}^{act}(\boldsymbol{z})$ and faults $\boldsymbol{f} \in \mathcal{N}_{\boldsymbol{f}}$ in each state $\boldsymbol{z} \in \mathcal{N}_{\boldsymbol{z}}$* exactly one *interconnection output $\bar{\boldsymbol{r}} \in \mathcal{N}_{\boldsymbol{r}}$ such that the relation*

$$\sum_{\boldsymbol{r}=1}^{T} \prod_{i=1}^{\nu} \chi_{\boldsymbol{F}_i^d}(r_i, z_i, v_i, \boldsymbol{k}_i^T \cdot \boldsymbol{r}, f_i) = \sum_{\boldsymbol{r}=\bar{\boldsymbol{r}}} \prod_{i=1}^{\nu} \chi_{\boldsymbol{F}_i^d}(\bar{r}_i, z_i, v_i, \boldsymbol{k}_i^T \cdot \bar{\boldsymbol{r}}, f_i) = 1 \tag{4.4}$$

holds. This interconnection output $\bar{\boldsymbol{r}}$ is called a fixed-point. ◇

Theorem 1.
The equivalent I/O-automaton $\tilde{\mathcal{D}}$ obtained by the composition rule (3.78) *from a deterministic asynchronous network of I/O-automata $\mathcal{DAN}$ is a deterministic I/O-automaton if and only if the $\mathcal{DAN}$ is well-posed.* □

Proof. Relation (4.4) can be transformed using the definition of the interconnection relation (3.44) to obtain:

$$\begin{aligned}
\forall (z_i, v_i, f_i)\ \exists \bar{\boldsymbol{r}} : & \sum_{\boldsymbol{r}=\bar{\boldsymbol{r}}} \prod_{i=1}^{\nu} \chi_{\boldsymbol{F}_i^d}(\bar{r}_i, z_i, v_i, \boldsymbol{k}_i^T \cdot \bar{\boldsymbol{r}}, f_i) = 1 \\
& \Leftrightarrow \sum_{\boldsymbol{r}=1}^{T} \prod_{i=1}^{\nu} \chi_{\boldsymbol{F}_i^d}(r_i, z_i, \boldsymbol{v}_i, \boldsymbol{k}_i^T \cdot \boldsymbol{r}, f_i) = 1 \\
& \Leftrightarrow \sum_{\boldsymbol{r}=1}^{T} \prod_{i=1}^{\nu} \sum_{z_i'=1}^{N_i} \sum_{w_i=1}^{R_i} \chi_{\boldsymbol{L}_i^d}(z_i', w_i, r_i, z_i, v_i, \boldsymbol{k}_i^T \cdot \boldsymbol{r}, f_i) = 1.
\end{aligned}$$

In the next step, the relation $\prod_{i=1}^{\nu} \sum_{z_i'}^{N_i} \sum_{w_i}^{R_i} \widehat{=} \sum_{\boldsymbol{z}'=1}^{N} \sum_{\boldsymbol{w}=1}^{R} \prod_{i=1}^{\nu}$ is used with changing the order of summation to get

$$\Leftrightarrow \sum_{\boldsymbol{z}'=1}^{N} \sum_{\boldsymbol{w}=1}^{R} \sum_{\boldsymbol{r}=1}^{T} \prod_{i=1}^{\nu} \chi_{\boldsymbol{L}_i^d}(z_i', w_i, r_i, z_i, v_i, \boldsymbol{k}_i^T \cdot \boldsymbol{r}, f_i) = 1.$$

Finally, the composition law (3.78) is used to wind up with the extension of eqn. (3.23) to deterministic I/O-automata without coupling signals but multiple control signals

$$\Leftrightarrow \sum_{\boldsymbol{z}'=1}^{N} \sum_{\boldsymbol{w}=1}^{R} \chi_{\tilde{\boldsymbol{L}}^d}(\boldsymbol{z}', \boldsymbol{w}, \boldsymbol{z}, \boldsymbol{v}, \boldsymbol{f}) = 1$$

which proves Theorem 1. ■

Corollary 1. *The feedback connection leads in general to a nondeterministic I/O-automaton, if the relation* $\exists(v_i, f_i) : \sum_{\boldsymbol{r}=1}^{T} \prod_{i=1}^{\nu} \chi_{\boldsymbol{F}_i^d}(r_i, z_i, \boldsymbol{v}_i, \boldsymbol{k}_i^T \cdot \boldsymbol{r}, f_i) \geq 1$ *is satisfied. The resulting I/O-automaton is called* not alive, *if* $\exists(v_i, f_i) : \sum_{\boldsymbol{r}=1}^{T} \prod_{i=1}^{\nu} \chi_{\boldsymbol{F}_i^d}(r_i, z_i, \boldsymbol{v}_i, \boldsymbol{k}_i^T \cdot \boldsymbol{r}, f_i) = 0$ *holds.* ◇

If the first condition of Corollary 1 holds, the special case of *weak determinism* has to be considered. It allows for several fixed-points $\bar{\boldsymbol{r}}$ solving eqn. (4.4) but demands a deterministic behavior according to the control input and output, i.e. there exist several internal signal ways that are not seen from outside the system. Hence, the network behaves internally nondeterministically whereas the I/O-automaton that results from composition is deterministic. This case has been studied in [11] and published in [6].

Definition 10. *(Weakly deterministic composition) The deterministic asynchronous network of I/O-automata* $\mathcal{DAN}$ (3.69) *is called* weakly well-posed, *if and only if the condition* (4.3) *holds and there exists for all activated control inputs* $\boldsymbol{v} \in \tilde{\mathcal{N}}_{\boldsymbol{v}}^{\text{act}}(\boldsymbol{z})$ *and faults* $\boldsymbol{f} \in \mathcal{N}_{\boldsymbol{f}}$ *in each state* $\boldsymbol{z} \in \mathcal{N}_{\boldsymbol{z}}$ at least one *interconnection output* $\bar{\boldsymbol{r}} \in \mathcal{N}_{\boldsymbol{r}}$ *such that the relation*

$$\sum_{\boldsymbol{z}'=1}^{N} \sum_{\boldsymbol{w}=1}^{R} \bigvee_{\boldsymbol{r}=1}^{T} \bigwedge_{i=1}^{\nu} \chi_{\boldsymbol{L}_i^d}(z_i', w_i, r_i, z_i, v_i, \boldsymbol{k}_i^T \cdot \boldsymbol{r}, f_i) = \tag{4.5}$$

$$\sum_{\boldsymbol{z}'=1}^{N} \sum_{\boldsymbol{w}=1}^{R} \bigvee_{\bar{\boldsymbol{r}}=1}^{\tilde{T}} \bigwedge_{i=1}^{\nu} \chi_{\boldsymbol{L}_i^d}(z_i', w_i, \bar{r}_i, z_i, v_i, \boldsymbol{k}_i^T \cdot \bar{\boldsymbol{r}}, f_i) = 1.$$

is satisfied. These interconnection outputs $\bar{\boldsymbol{r}} \in \mathcal{N}_{\bar{\boldsymbol{r}}} = \{1, \ldots, \tilde{T}\}$ *are called* fixed-points. ◇

> **Theorem 2.**
> *The equivalent I/O-automaton* $\tilde{\mathcal{D}}$ *of a system consisting of a deterministic asynchronous network of I/O-automata* $\mathcal{DAN}$ *is a deterministic I/O-automaton, if and only if it is weakly well-posed.* □

Proof. Condition (4.5) is stated on the behavioral relations opposed to condition (4.4) that is given in terms of the interconnection relations. The relation

$$\forall(z_i, v_i, f_i)\ \exists \bar{\boldsymbol{r}} : \bigvee_{\bar{\boldsymbol{r}}=1}^{\tilde{T}} \bigwedge_{i=1}^{\nu} \chi_{\boldsymbol{L}_i^d}(z_i', w_i, \bar{r}_i, z_i, v_i, \boldsymbol{k}_i^T \cdot \bar{\boldsymbol{r}}, f_i) = 1$$

$$\Leftrightarrow \bigvee_{\boldsymbol{r}=1}^{T} \bigwedge_{i=1}^{\nu} \chi_{\boldsymbol{L}_i^d}(z_i', w_i, r_i, z_i, v_i, \boldsymbol{k}_i^T \cdot \boldsymbol{r}, f_i) = 1$$

holds for the inner part of eqn. (4.5), if there exists at least one fixed-point $\bar{\boldsymbol{r}} \in \mathcal{N}_{\boldsymbol{r}}$. Thus, it can be transformed to

$$\Leftrightarrow \chi_{\boldsymbol{L}^d}(\boldsymbol{z}', \boldsymbol{w}, \boldsymbol{z}, \boldsymbol{v}, \boldsymbol{f}) = 1.$$

Inserting this result into eqn. (4.5) yields the above described extension of eqn. (3.23)

$$\Leftrightarrow \sum_{\boldsymbol{z}'=1}^{N} \sum_{\boldsymbol{w}=1}^{R} \chi_{\boldsymbol{L}^d}(z_i', w_i, z_i, v_i, f_i) = 1$$

and proves Theorem 2. ■

Nondeterministic behavior. In this case, the formal composition for nondeterministic networks of I/O-automata to equivalent nondeterministic I/O-automata without coupling signals given by eqn. (3.79) is used.

Definition 11. *(Nondeterministic composition) The nondeterministic asynchronous network of I/O-automata $\mathcal{NAN}$* (3.72) *is called* well-posed, *if and only if the following conditions hold.*

1. *The relation*

$$\mathcal{N}_{r_i} \subseteq \mathcal{N}_{s_j} \tag{4.6}$$

holds for all interconnection inputs and interconnection outputs that are related by the coupling model $\boldsymbol{K}$ to one another.

2. *There exists for all activated control inputs $\boldsymbol{v} \in \tilde{\mathcal{N}}_{\boldsymbol{v}}^{act}(\boldsymbol{z})$ and faults $\boldsymbol{f} \in \mathcal{N}_{\boldsymbol{f}}$ in each state $\boldsymbol{z} \in \mathcal{N}_{\boldsymbol{z}}$* at least one *coupling output $\bar{\boldsymbol{r}} \in \mathcal{N}_{\boldsymbol{r}}$ such that the relation*

$$\bigvee_{\boldsymbol{r}=1}^{T} \bigwedge_{i=1}^{\nu} \chi_{\boldsymbol{F}_i^n}(r_i, z_i, v_i, \boldsymbol{k}_i^T \cdot \boldsymbol{r}, f_i) = \bigvee_{\bar{\boldsymbol{r}}=1}^{\tilde{T}} \bigwedge_{i=1}^{\nu} \chi_{\boldsymbol{F}_i^n}(\bar{r}_i, z_i, v_i, \boldsymbol{k}_i^T \cdot \bar{\boldsymbol{r}}, f_i) = 1 \tag{4.7}$$

holds. These interconnection outputs $\bar{\boldsymbol{r}} \in \mathcal{N}_{\bar{\boldsymbol{r}}} = \{1, \ldots, \tilde{T}\}$ are called fixed-points. ◊

Theorem 3.
The equivalent I/O-automaton $\tilde{\mathcal{N}}$ of a system consisting of a nondeterministic asynchronous network of I/O-automata $\mathcal{NAN}$ is a nondeterministic I/O-automaton, if and only if the $\mathcal{NAN}$ is well-posed. □

Proof. Relation (4.7) can be transformed using the definition of the interconnection relation (3.47) to obtain:

$$\forall(z_i, v_i, f_i)\ \exists\bar{\boldsymbol{r}} : \bigvee_{\bar{\boldsymbol{r}}=1}^{\bar{T}} \bigwedge_{i=1}^{\nu} \chi_{\boldsymbol{F}_i^n}(\bar{r}_i, z_i, v_i, \boldsymbol{k}_i^T \cdot \bar{\boldsymbol{r}}, f_i) = 1$$

$$\Leftrightarrow \bigvee_{\boldsymbol{r}=1}^{T} \bigwedge_{i=1}^{\nu} \chi_{\boldsymbol{F}_i^n}(r_i, z_i, v_i, \boldsymbol{k}_i^T \cdot \boldsymbol{r}, f_i) = 1$$

$$\Leftrightarrow \bigvee_{\boldsymbol{r}=1}^{T} \bigwedge_{i=1}^{\nu} \bigvee_{z_i'=1}^{N_i} \bigvee_{w_i=1}^{R_i} \chi_{\boldsymbol{L}_i^n}(z_i', w_i, r_i, z_i, v_i, \boldsymbol{k}_i^T \cdot \boldsymbol{r}, f_i) = 1.$$

In the next step, the relation $\bigwedge_{i=1}^{\nu} \bigvee_{z_i'}^{N_i} \bigvee_{w_i}^{R_i} \widehat{=} \bigvee_{\boldsymbol{z}'=1}^{N} \bigvee_{\boldsymbol{w}=1}^{R} \bigwedge_{i=1}^{\nu}$ is used to get

$$\Leftrightarrow \bigvee_{\boldsymbol{z}'=1}^{N} \bigvee_{\boldsymbol{w}=1}^{R} \bigvee_{\boldsymbol{r}=1}^{T} \bigwedge_{i=1}^{\nu} \chi_{\boldsymbol{L}_i^n}(z_i', w_i, r_i, z_i, v_i, \boldsymbol{k}_i^T \cdot \boldsymbol{r}, f_i) = 1.$$

Finally, the composition law (3.79) is used to wind up with the extension of eqn. (3.40) to I/O-automata without coupling signals but multiple control signals

$$\Leftrightarrow \bigvee_{\boldsymbol{z}'=1}^{N} \bigvee_{\boldsymbol{w}=1}^{R} \chi_{\tilde{\boldsymbol{L}}^n}(\boldsymbol{z}', \boldsymbol{w}, \boldsymbol{z}, \boldsymbol{v}, \boldsymbol{f}) = 1$$

which proves Theorem 3. ■

Corollary 2. *The I/O-automaton that results from composing the network is called* not alive, *if* $\bigvee_{\boldsymbol{r}} \bigwedge_{i=1}^{\nu} \chi_{\boldsymbol{L}_i^n}(z_i', w_i, r_i, z_i, v_i, \boldsymbol{k}_i^T \cdot \boldsymbol{r}, f_i) = 0$ *holds.* ◇

Example 4 *Composition of two I/O-automata with asynchronous behavior*

The new modeling formalism is illustrated by a system consisting of two components C_1 *and* C_2 *whose models* $\mathcal{N}_1$ *and* $\mathcal{N}_2$ *are given in terms of their automaton graphs shown in Fig. 4.1. For simplicity, faults are not considered in this example such that a state transition is labeled by* $v_i/w_i/s_i/r_i$. *The corresponding interconnection relations* $\boldsymbol{F}_1$ *and* $\boldsymbol{F}_2$ *are given in Tab. 4.1.*

The interconnection signals are related by

$$s_1 = r_2 \qquad\qquad s_2 = r_1 \tag{4.8}$$

such that the system represents a feedback connection. It has to be checked whether this system model is well-posed or not. The difference in the deterministic and non-deterministic composition is explained in the following.

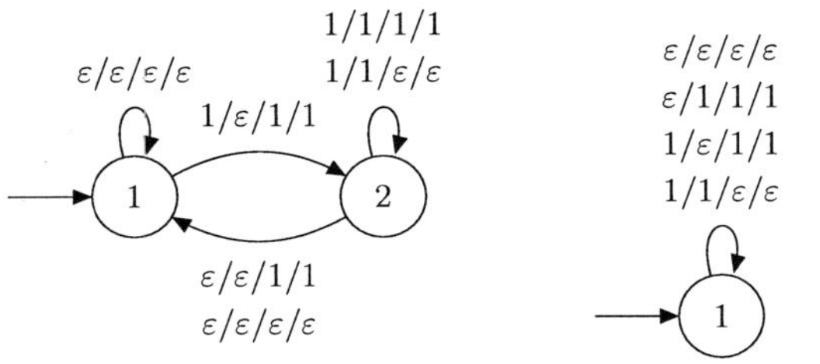

(a) Automaton $\mathcal{N}_1$ (b) Automaton $\mathcal{N}_2$

Figure 4.1: Automaton graph of the component models

Table 4.1: Interconnection relations

(a) $\boldsymbol{F}_1$

r_1	z_1	v_1	s_1
ε	1	ε	ε
1	1	1	1
ε	2	ε	ε
ε	2	1	ε
1	2	ε	1
1	2	1	1

(b) $\boldsymbol{F}_2$

r_2	z_2	v_2	s_2
ε	1	ε	ε
1	1	ε	1
1	1	1	1
ε	1	1	ε

Deterministic composition. *According to Definition 9, relation (4.3) holds for this example. Equation (4.4) to be checked in each state* $\boldsymbol{z}$

$$\boldsymbol{z} = (z_1, z_2)^T \in \mathcal{N}_{\boldsymbol{z}} = \left\{(1,1)^T, (2,1)^T\right\} \tag{4.9}$$

for all activated control inputs $\boldsymbol{v}$

$$\boldsymbol{v} = \begin{pmatrix} v_1 \\ v_2 \end{pmatrix} \in \mathcal{N}_{\boldsymbol{v}} = \left\{ \begin{pmatrix} \varepsilon \\ \varepsilon \end{pmatrix}, \begin{pmatrix} 1 \\ \varepsilon \end{pmatrix}, \begin{pmatrix} \varepsilon \\ 1 \end{pmatrix}, \begin{pmatrix} 1 \\ 1 \end{pmatrix} \right\} \tag{4.10}$$

by summing the product of the interconnection relations $\boldsymbol{F}_1$ *and* $\boldsymbol{F}_2$ *for all interconnection outputs* $\boldsymbol{r}$

$$\boldsymbol{r} = \begin{pmatrix} r_1 \\ r_2 \end{pmatrix} \in \mathcal{N}_{\boldsymbol{r}} = \left\{ \begin{pmatrix} \varepsilon \\ \varepsilon \end{pmatrix}, \begin{pmatrix} 1 \\ \varepsilon \end{pmatrix}, \begin{pmatrix} \varepsilon \\ 1 \end{pmatrix}, \begin{pmatrix} 1 \\ 1 \end{pmatrix} \right\}. \tag{4.11}$$

This summation is shown in Table 4.2. It can be seen from Tab. 4.2(a) that condition (4.4) holds only in state $\boldsymbol{z} = (1,1)^T$ *but not in state* $\boldsymbol{z} = (2,1)^T$ *because the summation is equal to 2 for all control inputs (Table 4.2(b)). Hence, the overall system is not well-posed according to Theorem 1.*

A well-posed system can be obtained by removing in the automaton graph of $\mathcal{N}_1$ *in state 2 the transitions labeled by* $\varepsilon/\varepsilon/\varepsilon/\varepsilon$ *and* $1/1/1/1$*. Then, there is exactly*

Table 4.2: Checking (4.4) for $\mathcal{N}_1$ and $\mathcal{N}_2$

(a) in $\boldsymbol{z} = (1,1)^T$

$(v_1, v_2)^T$	$\boldsymbol{F}_1(\bullet) \cdot \boldsymbol{F}_2(\bullet)$ for $\boldsymbol{r} =$				$\sum_{\boldsymbol{r}} \boldsymbol{F}_1(\bullet) \cdot \boldsymbol{F}_2(\bullet)$
	$(\varepsilon,\varepsilon)^T$	$(1,\varepsilon)^T$	$(\varepsilon,1)^T$	$(1,1)^T$	
$(\varepsilon,\varepsilon)^T$	$1 \cdot 1$	$0 \cdot 0$	$0 \cdot 0$	$0 \cdot 1$	1
$(1,\varepsilon)^T$	$0 \cdot 1$	$0 \cdot 0$	$0 \cdot 0$	$1 \cdot 1$	1
$(\varepsilon,1)^T$	$1 \cdot 1$	$0 \cdot 0$	$0 \cdot 0$	$0 \cdot 1$	1
$(1,1)^T$	$0 \cdot 1$	$0 \cdot 0$	$0 \cdot 0$	$1 \cdot 1$	1

(b) in $\boldsymbol{z} = (2,1)^T$

$(v_1, v_2)^T$	$\boldsymbol{F}_1(\bullet) \cdot \boldsymbol{F}_2(\bullet)$ for $\boldsymbol{r} =$				$\sum_{\boldsymbol{r}} \boldsymbol{F}_1(\bullet) \cdot \boldsymbol{F}_2(\bullet)$
	$(\varepsilon,\varepsilon)^T$	$(1,\varepsilon)^T$	$(\varepsilon,1)^T$	$(1,1)^T$	
$(\varepsilon,\varepsilon)^T$	$1 \cdot 1$	$0 \cdot 0$	$0 \cdot 0$	$1 \cdot 1$	2
$(1,\varepsilon)^T$	$1 \cdot 1$	$0 \cdot 0$	$0 \cdot 0$	$1 \cdot 1$	2
$(\varepsilon,1)^T$	$1 \cdot 1$	$0 \cdot 0$	$0 \cdot 0$	$1 \cdot 1$	2
$(1,1)^T$	$1 \cdot 1$	$0 \cdot 0$	$0 \cdot 0$	$1 \cdot 1$	2

one combination of interconnection signals solving eqn. (4.4) *for the state* $(2,1)^T$ *such that the system is well-posed. The automaton graph of the composition of the reduced component models is depicted in Fig. 4.2, where a directed edge is labeled by* $\boldsymbol{v}/\boldsymbol{w}$.

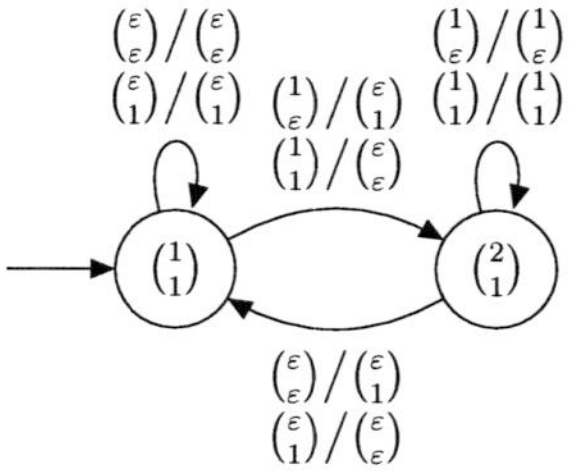

Figure 4.2: Automaton graph of the composed I/O-automaton $\mathcal{N}^{comp}$

The state transitions of the overall system can be separated into synchronous and asynchronous ones.

1. **Synchronous state transitions** *occur, if both components get non-vanishing control input symbols. In the example, this is e.g. the case for a transition from state* $\boldsymbol{z} = (1,1)^T$ *to the successor state* $\boldsymbol{z}' = (2,1)^T$ *for the control input* $\boldsymbol{v} = (1,1)^T$ *and the generated control output* $\boldsymbol{w} = (\varepsilon,\varepsilon)^T$. *Another possibility is that there is no interaction due to non-vanishing input symbols, e.g. the self loop at state* $\boldsymbol{z} = (2,1)^T$ *for the control input* $\boldsymbol{v} = (1,1)^T$ *and the control output*

$\boldsymbol{w} = (\varepsilon/\varepsilon)^T$. *It is also possible that one component gets a non-vanishing control input while the other one gets the empty symbol. This case is given by the state transition from state* $\boldsymbol{z} = (1,1)^T$ *to the successor state* $\boldsymbol{z}' = (2,1)^T$. *Even though the control input* $\boldsymbol{v} = (1,\varepsilon)^T$ *has one vanishing element for the first component, the control output* $\boldsymbol{w} = (\varepsilon, 1)^T$ *can have the non-vanishing element for the second one.*

2. *An* **asynchronous state transition** *can be performed, if one component gets a non-vanishing control input while the other one gets an* ε *and there is no interaction between the components. This is for example the case for the self loop at state* $\boldsymbol{z} = (2,1)^T$ *and for the control input* $\boldsymbol{v} = (1,\varepsilon)^T$ *and for the control output* $\boldsymbol{w} = (1,\varepsilon)^T$.

Weakly deterministic composition. *Condition (4.4) holds only in state* $\boldsymbol{z} = (1,1)^T$ *as stated above. It can be seen from Table 4.2(b) that there exist two combinations of interconnection outputs solving eqn. (4.4) for all activated control inputs in state* $\boldsymbol{z} = (2,1)^T$. *The condition for weak determinism is investigated in Table 4.3. The I/O-automaton* $\mathcal{N}_1$ *behaves deterministic in both cases but the I/O-automaton* $\mathcal{N}_2$ *does not do so due to the different control outputs that are generated according to the different interconnection outputs. Weak determinism can be achieved either by relabeling the transitions* $\varepsilon/\varepsilon/\varepsilon/\varepsilon$ *and* $1/1/\varepsilon/\varepsilon$ *with* $\varepsilon/1/\varepsilon/\varepsilon$ *and* $1/\varepsilon/\varepsilon/\varepsilon$ *or the transitions* $\varepsilon/1/1/1$ *and* $1/\varepsilon/1/1$ *with* $\varepsilon/\varepsilon/1/1$ *and* $1/1/1/1$.

Table 4.3: Checking (4.5) for $\mathcal{N}_1$ and $\mathcal{N}_2$ in the case of multiple solutions

$(v_1, v_2)^T$	$\boldsymbol{r} = (\varepsilon/\varepsilon)^T$		$\boldsymbol{r} = (1/1)^T$	
	$(z_1, w_1) =$	(z_2, w_2)	$(z_1, w_1) =$	(z_2, w_2)
$(\varepsilon,\varepsilon)^T$	$(1,\varepsilon)$	$(1,\varepsilon)$	$(1,\varepsilon)$	$(1,1)$
$(1,\varepsilon)^T$	$(2,\varepsilon)$	$(1,\varepsilon)$	$(2,\varepsilon)$	$(1,1)$
$(\varepsilon,1)^T$	$(1,\varepsilon)$	$(1,1)$	$(1,\varepsilon)$	$(1,\varepsilon)$
$(1,1)^T$	$(2,\varepsilon)$	$(1,1)$	$(2,\varepsilon)$	$(1,\varepsilon)$

Nondeterministic composition. *Relation (4.7) holds because relation (4.4) is satisfied. The check of eqn. (4.6) is obvious considering the previously derived results. The I/O-automaton resulting from the nondeterministic composition is shown in Fig. 4.3. The nondeterminism arises only in state* $\boldsymbol{z} = (2,1)^T$. *It is due to the different control outputs that are generated for one and the same control input. This is e.g. the case for the transition from state* $\boldsymbol{z} = (2,1)^T$ *to itself for the control input* $\boldsymbol{v} = (1,\varepsilon)^T$ *which generates the two control outputs* $\boldsymbol{w} = (1,\varepsilon)^T$ *and* $\boldsymbol{w} = (1,1)^T$.

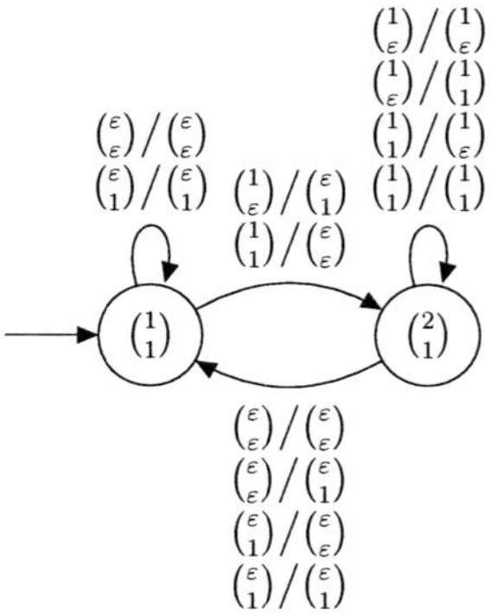

Figure 4.3: Automaton graph of the composed I/O-automaton $\mathcal{N}^{comp}$

4.2 Autonomy of asynchronous networks of I/O-automata

Autonomy corresponds to case where interactions have no effect on the component behavior. Two different types will be investigated in this section. Structural autonomy goes along with the analysis of the coupling models and is considered in Section 4.2.1. It does not change during operation. Thus, the system can be divided permanently into subsystems which can be investigated independently of each other. The more interesting case of state-dependent autonomy is analyzed in Section 4.2.2 based on the behavioral relation of the components. It results in temporal autonomy of a possibly changing set of components such that the system can be divided occasionally but not always into subsystems.

4.2.1 Structural autonomy

Structural autonomy corresponds to the analysis of the coupling model $\boldsymbol{K}$. It is inspired from graph theory [37, 112] and the search for sets of strongly connected vertices (nodes) [70]. To apply the known analysis methods, the network of I/O-automata is interpreted as the *interconnection graph*

$$\mathcal{G} = (\mathcal{V}, \mathcal{A}) \tag{4.12}$$

where the set of vertices $\mathcal{V}$ corresponds to the components C_i of the system and the set of directed arcs (edges) $\mathcal{A}$ to the interconnections between the components. An arc $a_i \in \mathcal{A}$ exists, if at least one interconnection input of the component C_i is connected with an interconnection output of the component C_j. Thus, a_i starts at C_j and ends at C_i. The corresponding element A_{ij} of the associated matrix $\boldsymbol{A}$ of the graph $\mathcal{G}$ has the value "1". This matrix corresponds in the case of single interconnection signals directly to the coupling model $\boldsymbol{K}$. A path is defined in the interconnection graph as a sequence of arcs where the end vertex of each arc corresponds to the start vertex of the next arc [70].

Definition 12. *(Strongly connected vertices) Two vertices n_i and $n_j \in \mathcal{V}$ are said to be strongly connected, if and only if there exists a path from vertex n_i to vertex n_j and a path from n_j to n_i.* ◊

The graph $\mathcal{G}$ is finite because the set of vertices corresponds to the finite set of components of the system. Hence, the longest simple path between two vertices has the length $\nu - 1$. The reachability of one vertex from another one can be obtained by the matrix

$$\bar{\boldsymbol{A}} = \sum_{i=1}^{\nu-1} \boldsymbol{A}^i. \tag{4.13}$$

A path exists between two vertices n_i and n_j, if the element $\bar{A}_{ij}$ is unequal to zero.

Important rules from graph theory. The following results are known from graph theory [70, 112] and can be obtained by the analysis of the matrix $\bar{\boldsymbol{A}}$:

1. Two vertices n_i and n_j are strongly connected, if $\bar{A}_{ij} = \bar{A}_{ji} \neq 0$.
2. Two vertices n_i and n_j are not connected, if $\bar{A}_{ij} = \bar{A}_{ji} = 0$.
3. The set of vertices can be partitioned into not connected, disjoint sets of strongly connected vertices

   $$\mathcal{V} = \mathcal{V}_1 \cup \mathcal{V}_2 \cup \ldots \cup \mathcal{V}_\alpha, \qquad \mathcal{V}_{\tilde{i}} \cap \mathcal{V}_{\tilde{j}} = \emptyset \text{ for } \tilde{i} \neq \tilde{j} \tag{4.14}$$

   with $1 \leq \alpha \leq \nu$, if $\bar{A} = \bar{A}^T$ and $\bar{A}_{ij} = \bar{A}_{ji} = 0$ holds for all $i \in \tilde{i}$ and $j \in \tilde{j}$ where $\tilde{i}$ and $\tilde{j}$ have been defined on page 41. A set $\mathcal{V}_i$ is called an equivalence class.
4. All vertices are strongly connected, if $\bar{A}_{ij} = \bar{A}_{ji} \neq 0$ holds for all $i, j \in \{1, \ldots, \nu\}$ and $i \neq j$. This case can be split up in the following way:
   a. The connection of all vertices is given as one loop of length ν, if additionally $\bar{A}_{ii} = 0$ holds for all $i \in \{1, \ldots, \nu\}$.
   b. A vertex n_i is connected with itself in a loop of length less than ν, if $\bar{A}_{ii} \neq 0$ is given for the corresponding element.
5. All vertices n_i and n_j are only connected with themselves, if $\bar{A}_{ij} = \bar{A}_{ji} = 0$ holds for all $i, j \in \{1, \ldots, \nu\}, i \neq j$ and $\bar{A}_{ii} \neq 0$ for all $i \in \{1, \ldots, \nu\}$.
6. Two vertices n_i and n_j are connected in series, if either $\bar{A}_{ij} \neq$ or $\bar{A}_{ji} \neq 0$ holds but not both.

7. Two sets of strongly connected vertices $\mathcal{V}_{\tilde{i}}$ and $\mathcal{V}_{\tilde{j}}$ are connected in series, if either $\bar{A}_{ij} \neq 0$ or $\bar{A}_{ji} \neq 0$ but not both holds for all $i \in \tilde{i}$ and $j \in \tilde{j}$. ◇

Definition 13. *(Structural analysis of the coupling model) The following definitions are made based on the analysis of the coupling model:*

1. *Two components of a system are said to be* structurally autonomous *to one another, if they belong to different equivalence classes $\mathcal{V}_i$ given by eqn.* (4.14) *which are connected via a path in neither direction.*
2. *A system is said to be* totally heteronomous, *if there exists only one equivalence class.*
3. *A system is said to be* totally autonomous, *if there exist ν equivalence classes that are connected with one another via a path.*
4. *Two groups of components are said to be* connected in series, *if there exists a path from the equivalence class $\mathcal{V}_i$, which corresponds with the first group, to the equivalence class $\mathcal{V}_j$, which corresponds with the second group, but not vice versa.* ◇

Consequences from structural autonomy. The following consequences regarding the analysis of an asynchronous network of I/O-automata can be drawn from Definition 13:

1. The analysis of structurally autonomous components can be carried out separately according to the equivalence classes because there is no interaction between the components and by this no mutual influence.
2. A total heteronomous system must be analyzed holistically. Thus, the analysis cannot be simplified.
3. Totally autonomous systems can be analyzed component-wise because there is no interaction between any pair of components.
4. The analysis of two equivalence classes $\mathcal{V}_i$ and $\mathcal{V}_j$ that are connected in series can be split up such that $\mathcal{V}_i$ can be considered separately whereas $\mathcal{V}_j$ must be analyzed in conjunction with the influence of $\mathcal{V}_i$.

According to these statements, the system under consideration is structurally analyzed in the first place and split up into its equivalence classes. The analysis of each equivalence class is carried out independently from the remaining equivalence classes as long as there is no interaction with the other equivalence classes. Thus, it is assumed for the remainder of this thesis that the system under consideration *cannot be separated any further* and must be analyzed on the whole.

Example 5 *Examples for structural autonomy in interconnected systems*

Four different examples are given according to Definition 13. The first case corresponds to structural autonomy shown in Fig. 4.4. The associated matrix $\boldsymbol{A}$ *of the interconnection graph and the matrix* $\bar{\boldsymbol{A}}$ *are given as*

$$\boldsymbol{A} = \begin{pmatrix} 0 & 1 & 0 \\ 1 & 0 & 0 \\ 0 & 0 & 1 \end{pmatrix}, \qquad \bar{\boldsymbol{A}} = \begin{pmatrix} 1 & 1 & 0 \\ 1 & 1 & 0 \\ 0 & 0 & 2 \end{pmatrix}.$$

Thus, $\bar{\boldsymbol{A}} = \bar{\boldsymbol{A}}^T$ *and* $\bar{A}_{13} = \bar{A}_{31} = 0 = \bar{A}_{23} = \bar{A}_{32}$ *holds which corresponds to the third rule of graph theory.*

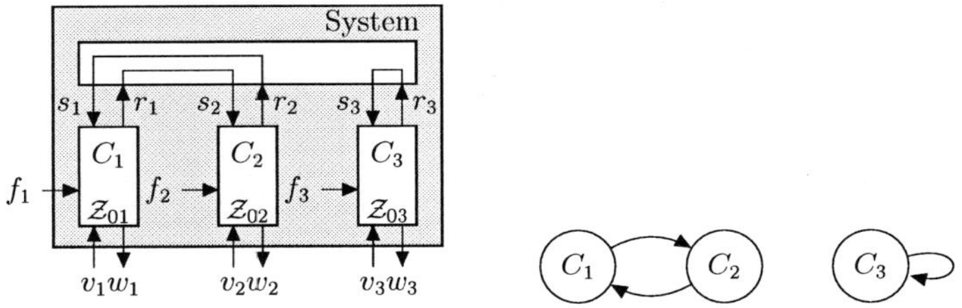

(a) Interconnected system (b) Interconnection graph

Figure 4.4: Structural autonomy in interconnected discrete-event systems

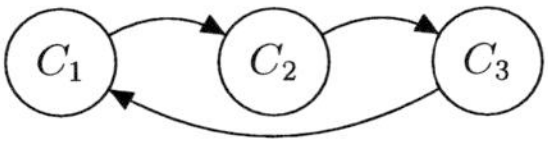

Figure 4.5: Interconnection graph of the system shown in Fig. 3.6(a)

The case of total heteronomy is depicted in Fig. 3.6(a). It can be seen from the analysis of the associated matrix $\boldsymbol{A}$ *of the interconnection graph given in Fig. 4.5 and the matrix* $\bar{\boldsymbol{A}}$ *that the conditions of the fourth rule of graph theory holds:*

$$\boldsymbol{A} = \begin{pmatrix} 0 & 0 & 1 \\ 1 & 0 & 0 \\ 0 & 1 & 0 \end{pmatrix}, \qquad \bar{\boldsymbol{A}} = \begin{pmatrix} 0 & 1 & 1 \\ 1 & 0 & 1 \\ 1 & 1 & 0 \end{pmatrix}.$$

In addition, as the elements of the main diagonal of $\bar{\boldsymbol{A}}$ *are equal to zero, there exist no simple paths from each component to itself.*

Next, the case of total autonomy is investigated. An example is given in Fig. 4.6. The associated matrix $\boldsymbol{A}$ *of the graph and the matrix* $\bar{\boldsymbol{A}}$ *are*

$$\boldsymbol{A} = \begin{pmatrix} 1 & 0 & 0 \\ 0 & 1 & 0 \\ 0 & 0 & 1 \end{pmatrix}, \qquad \bar{\boldsymbol{A}} = \begin{pmatrix} 2 & 0 & 0 \\ 0 & 2 & 0 \\ 0 & 0 & 2 \end{pmatrix}.$$

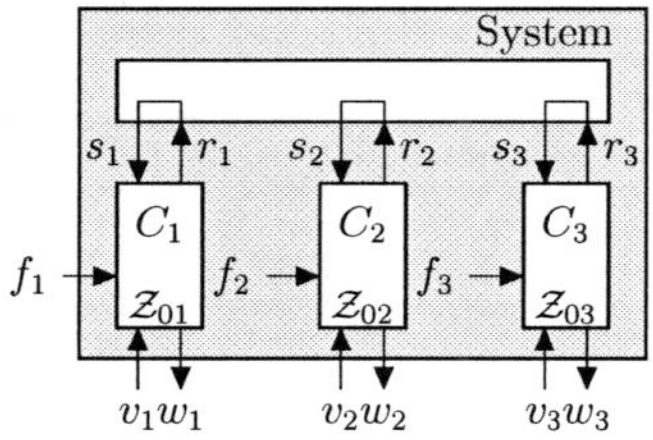

(a) Interconnected system

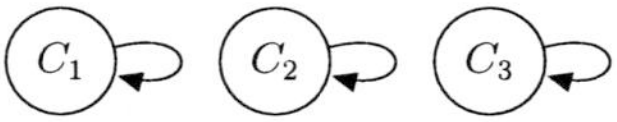

(b) Interconnection graph

Figure 4.6: Total autonomy in interconnected discrete-event systems

Thus, the conditions of the fifth rule of graph theory hold.

A series connection is given in Fig. 4.7 with its corresponding graph. It extends the first example by adding a connection from component C_2 to C_3. The associated matrix $\boldsymbol{A}$ and the matrix $\bar{\boldsymbol{A}}$ are specified as:

$$\boldsymbol{A} = \begin{pmatrix} 0 & 1 & 0 \\ 1 & 0 & 0 \\ 0 & 1 & 1 \end{pmatrix}, \qquad \bar{\boldsymbol{A}} = \begin{pmatrix} 1 & 1 & 0 \\ 1 & 1 & 0 \\ 1 & 2 & 2 \end{pmatrix}.$$

Obviously, $\bar{A}_{13} \neq \bar{A}_{31}$ and $\bar{A}_{23} \neq \bar{A}_{32}$ holds whereas $\bar{A}_{ij} = \bar{A}_{ij} \neq 0$ is true for the remaining elements. Hence, the strongly connected components C_1 and C_2 are connected in series with the strongly connected component C_3 which corresponds to the seventh rule of graph theory.

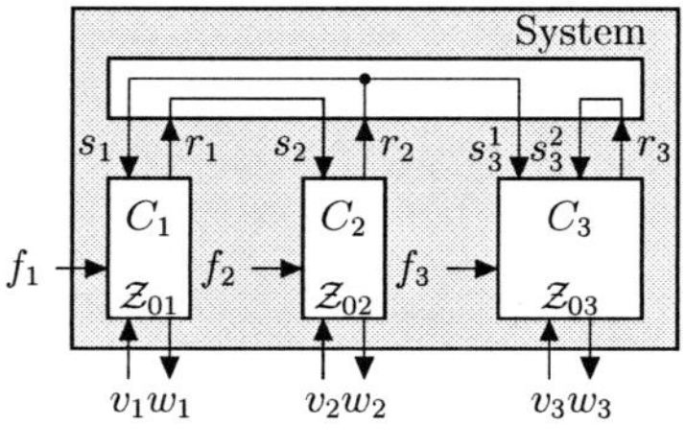

(a) Interconnected system

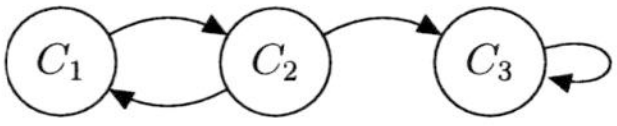

(b) Interconnection graph

Figure 4.7: Series connection in interconnected discrete-event systems

4.2.2 State-dependent autonomy

State-dependent autonomy is a property that a subsystem behaves independently of the state or state transition of the remaining system although it is coupled with these components via the coupling network. Hence, the values of the interconnection signals do not always have an influence on the behavior of the component and conditions must be given in terms of the *model* $\mathcal{N}_i$ of the component C_i.

Definition 14. *(State- and fault-dependent autonomy) The I/O-automaton $\mathcal{N}_i$ is said to behave*

1. autonomously in the state z_i for a specific combination of control input $\boldsymbol{v}_i$ and fault f_i, *if the characteristic function $\chi_{\boldsymbol{L}_i^n}$ is independent of the values of the coupling signals s_i and r_i in this state z_i for the fault f_i, the control input $\boldsymbol{v}_i$ and all activated interconnection inputs $\boldsymbol{s}_i \in \mathcal{N}_{\boldsymbol{s}_i}^{act}(z_i)$. Hence, there exists a characteristic function $\bar{\chi}_{\boldsymbol{L}_i^n}$ on the behavioral relation $\boldsymbol{L}_i^n$ such that the equation*

$$\chi_{\boldsymbol{L}_i^n}(z_i', \boldsymbol{w}_i, \boldsymbol{r}_i, z_i, \boldsymbol{v}_i, \boldsymbol{s}_i, f_i) = \bar{\chi}_{\boldsymbol{L}_i^n}(z_i', \boldsymbol{w}_i, z_i, \boldsymbol{v}_i, f_i) \tag{4.15}$$

holds for the characteristic function $\chi_{\boldsymbol{L}_i^n}$.

2. autonomously in the state z_i for a specific fault f_i, *if the characteristic function $\chi_{\boldsymbol{L}_i^n}$ is independent of the values of the coupling signals s_i and r_i in this state z_i for the fault f_i, all activated control inputs $\boldsymbol{v}_i \in \mathcal{N}_{\boldsymbol{v}_i}^{act}(z_i)$ and all activated interconnection inputs $\boldsymbol{s}_i \in \mathcal{N}_{\boldsymbol{s}_i}^{act}(z_i)$. Hence, there exists a characteristic function $\bar{\chi}_{\boldsymbol{L}_i^n}$ on the behavioral relation $\boldsymbol{L}_i^n$ such that the equation*

$$\chi_{\boldsymbol{L}_i^n}(z_i', \boldsymbol{w}_i, \boldsymbol{r}_i, z_i, \boldsymbol{v}_i, \boldsymbol{s}_i, f_i) = \bar{\chi}_{\boldsymbol{L}_i^n}(z_i', \boldsymbol{w}_i, z_i, f_i) \tag{4.16}$$

holds for the characteristic function $\chi_{\boldsymbol{L}_i^n}$. ◇

By the first case of Definition 14, the state transition does not depend on the activated interconnection input $\boldsymbol{s}_i$ and any interconnection output $\boldsymbol{r}_i$ but still on the fault f_i for a given control input $\boldsymbol{v}_i$. Thus, there may be combinations of control inputs and faults resulting in the autonomous motion of the component in the state z_i whereas the behavior of the component is influenced by other components for the remaining control inputs and faults. The set $\mathcal{Z}_i^{auto}(v_i, f_i)$ contains all states, control inputs and faults for which the autonomy condition (4.15) holds:

$$\begin{aligned}\mathcal{Z}_i^{auto}(v_i, f_i) = \{z_i \in \mathcal{N}_{z_i} | \exists\ (z_i', w_i, r_i) : \chi_{\boldsymbol{L}_i^n}(z_i', \boldsymbol{w}_i, \boldsymbol{r}_i, z_i, \boldsymbol{v}_i, \boldsymbol{s}_i, f_i) \\ = \bar{\chi}_{\boldsymbol{L}_i^n}(z_i', \boldsymbol{w}_i, z_i, \boldsymbol{v}_i, f_i)\ \forall\ \boldsymbol{s}_i \in \mathcal{N}_{\boldsymbol{s}_i}^{act}(z_i)\}.\end{aligned} \tag{4.17}$$

The second case of Definition 14 is more restrictive by demanding that eqn. (4.16) holds. Hence, the set $\mathcal{Z}_i^{auto}(f_i)$ is given as:

$$\begin{aligned}\mathcal{Z}_i^{auto}(f_i) = \{z_i \in \mathcal{N}_{z_i} | \exists\ (z_i', w_i, r_i) : \chi_{\boldsymbol{L}_i^n}(z_i', \boldsymbol{w}_i, \boldsymbol{r}_i, z_i, \boldsymbol{v}_i, \boldsymbol{s}_i, f_i) \\ = \bar{\chi}_{\boldsymbol{L}_i^n}(z_i', \boldsymbol{w}_i, z_i, f_i)\ \forall\ \boldsymbol{v}_i \in \mathcal{N}_{\boldsymbol{v}_i}^{act}(z_i), \boldsymbol{s}_i \in \mathcal{N}_{\boldsymbol{s}_i}^{act}(z_i)\}.\end{aligned} \tag{4.18}$$

Restricting eqns. (4.15) and (4.16) to hold for all faults, results in total autonomy of the component in a state z_i. This case is stated next.

Definition 15. *(Total state-dependent autonomy) The I/O-automaton $\mathcal{N}_i$ is said to behave*

1. totally autonomously in the state z_i for a specific control input $\boldsymbol{v}_i$, *if the characteristic function $\chi_{\boldsymbol{L}_i^n}$ is independent of the values of the coupling signals s_i and r_i in this state z_i for the control input $\boldsymbol{v}_i$, all activated interconnection inputs $\boldsymbol{s}_i \in \mathcal{N}_{\boldsymbol{s}_i}^{act}(z_i)$ and all faults $f_i \in \mathcal{N}_{f_i}$. Hence, there exists a characteristic function $\bar{\chi}_{\boldsymbol{L}_i^n}$ on the behavioral relation $\boldsymbol{L}_i^n$ such that the equation*

$$\chi_{\boldsymbol{L}_i^n}(z_i', \boldsymbol{w}_i, \boldsymbol{r}_i, z_i, \boldsymbol{v}_i, \boldsymbol{s}_i, f_i) = \bar{\chi}_{\boldsymbol{L}_i^n}(z_i', \boldsymbol{w}_i, z_i, \boldsymbol{v}_i) \tag{4.19}$$

holds for the characteristic function $\chi_{\boldsymbol{L}_i^n}$.

2. totally autonomously in the state z_i, *if the characteristic function $\chi_{\boldsymbol{L}_i^n}$ is independent of the values of the coupling signals s_i and r_i in this state z_i for all activated control inputs $\boldsymbol{v}_i \in \mathcal{N}_{\boldsymbol{v}_i}^{act}(z_i)$, all activated interconnection inputs $\boldsymbol{s}_i \in \mathcal{N}_{\boldsymbol{s}_i}^{act}(z_i)$ and all faults $f_i \in \mathcal{N}_{f_i}$. Hence, there exists a characteristic function $\bar{\chi}_{\boldsymbol{L}_i^n}$ on the behavioral relation $\boldsymbol{L}_i^n$ such that the equation*

$$\chi_{\boldsymbol{L}_i^n}(z_i', \boldsymbol{w}_i, \boldsymbol{r}_i, z_i, \boldsymbol{v}_i, \boldsymbol{s}_i, f_i) = \bar{\chi}_{\boldsymbol{L}_i^n}(z_i', \boldsymbol{w}_i, z_i) \tag{4.20}$$

holds for the characteristic function $\chi_{\boldsymbol{L}_i^n}$. $\diamond$

By the first case of Definition 15, the state transition does neither depend on the interconnection input $\boldsymbol{s}_i$ and output $\boldsymbol{r}_i$ nor on the fault f_i for a given control input $\boldsymbol{v}_i$. Thus, there may be control inputs that result in the total autonomous motion of the component in the state z_i whereas the behavior of the component is influenced by other components for the remaining control inputs. The resulting set $\mathcal{Z}_i^{auto}(v_i)$ contains all states and control inputs for which the autonomy condition (4.19) holds:

$$\begin{aligned} \mathcal{Z}_i^{auto}(v_i) = \{z_i \in \mathcal{N}_{z_i} | \exists\ (z_i', w_i, r_i) : \chi_{\boldsymbol{L}_i^n}(z_i', \boldsymbol{w}_i, \boldsymbol{r}_i, z_i, \boldsymbol{v}_i, \boldsymbol{s}_i, f_i) \\ = \bar{\chi}_{\boldsymbol{L}_i^n}(z_i', \boldsymbol{w}_i, z_i, \boldsymbol{v}_i)\ \forall\ \boldsymbol{s}_i \in \mathcal{N}_{\boldsymbol{s}_i}^{act}(z_i), f_i \in \mathcal{N}_{f_i}\}. \end{aligned} \tag{4.21}$$

The second case of Definition 15 is more restrictive by demanding that eqn. (4.20) holds. Hence, the set $\mathcal{Z}_i^{auto}$ is given as:

$$\begin{aligned} \mathcal{Z}_i^{auto} = \{z_i \in \mathcal{N}_{z_i} | \exists\ (z_i', w_i, r_i) : \chi_{\boldsymbol{L}_i^n}(z_i', \boldsymbol{w}_i, \boldsymbol{r}_i, z_i, \boldsymbol{v}_i, \boldsymbol{s}_i, f_i) \\ = \bar{\chi}_{\boldsymbol{L}_i^n}(z_i', \boldsymbol{w}_i, z_i)\ \forall\ \boldsymbol{v}_i \in \mathcal{N}_{\boldsymbol{v}_i}^{act}(z_i), \boldsymbol{s}_i \in \mathcal{N}_{\boldsymbol{s}_i}^{act}(z_i), f_i \in \mathcal{N}_{f_i}\}. \end{aligned} \tag{4.22}$$

If eqn. (4.20) holds additionally for *all* states $z_i \in \mathcal{N}_{z_i}$, the ith component acts completely independently of the remaining system. Hence, it can be analyzed completely separate from the other components. The more interesting case concerns subsystems where the autonomy conditions (4.15), (4.16), (4.19) and (4.20) are satisfied in some but not all states for some but not all faults in several states. The autonomy conditions are used in Chapters 6 and 7 for simplifications in diagnosis.

Example 6 *Example for state-dependent autonomy*

The automaton Table 4.4 is used to illustrate the different definitions of state-dependent autonomy. The transitions starting in state $z_i = 1$ are investigated first. The set of activated control inputs is given as $\mathcal{N}_{v_i}^{act}(1) = \{\varepsilon, 3\}$. It can be seen from the first two lines of the automaton table that state transitions to $z_i' = 1$ with the control output $w_i = 3$ are possible for all activated interconnection inputs and all faults. The third and the fourth line of the automaton table indicate that state transitions to $z_i' = 2$ with the control output $w_i = \varepsilon$ are possible for the control input $v_i = \varepsilon$, all activated interconnection inputs and all faults. Hence, the I/O-automaton is totally autonomously in the state $z_i = 1$ for the control input $v_i = \varepsilon$ and for the control input $v_i = 3$ respectively but it is not totally autonomously because different successor states and control outputs are possible for different control inputs.

Table 4.4: Exemplary automaton table

z_i'	w_i	r_i	z_i	v_i	s_i	f_i
1	3	ε	1	3	ε	0
1	3	ε	1	3	ε	1
2	ε	1	1	ε	3	0
2	ε	3	1	ε	1	1
1	1	2	2	1	1	0
1	1	1	2	1	2	0
2	1	2	2	1	1	1

The transitions starting in state $z_i = 2$ are analyzed next. The fifth and the sixth line of the automaton table show that transitions to the state $z_i' = 1$ with the control output $w_i = 1$ are possible for the control input $v_i = 1$ and the fault $f_i = 0$. The last line describes a transition to the state $z_i' = 2$ with the control output $w_i = 1$ for the control input $v_i = 1$ and the fault $f_i = 1$. Hence, the I/O-automaton is autonomously in the state $z_i = 1$ for the combination $v_i = 1$ and $f_i = 0$. It is additionally autonomous for the combination $v_i = 1$ and $f_i = 0$ but it behaves not autonomously for the control input $v_i = 1$ because different successor states are possible for different faults.

4.3 Simulation algorithms

Methods for the simulation of nondeterministic asynchronous networks of I/O-automata are presented in this section. The basic concept is explained in Section 4.3.1 using a single I/O-automaton. The extension to networks of I/O-automata is given in Section 4.3.2 using online composition of the behavioral relation to retain computability. These algorithms have been implemented and tested in MATLAB in the Diploma thesis [11] and published in [6] for the deterministic case. The extension to nondeterministic asynchronous networks of I/O-automata is considered in this section based on the simulation algorithms for stochastic networks of I/O-automata presented in [84] which relies on [113].

4.3.1 Behavior of single nondeterministic I/O-automata

The task of simulation is to calculate the sequences of outputs and states of a system based on a system model for a given sequence of inputs and a set of initial states (Fig. 4.8). The approaches presented here give a solution to all possible scenarios, i.e. combinations of inputs and initial states, simultaneously by calculating the possibility of all combinations of states and outputs for a given time k. The simulation algorithm for a single nondeterministic I/O-automaton is derived in two steps. First, an autonomous automaton is considered. The obtained results are than extended to I/O-automata with inputs and outputs [84, 113].

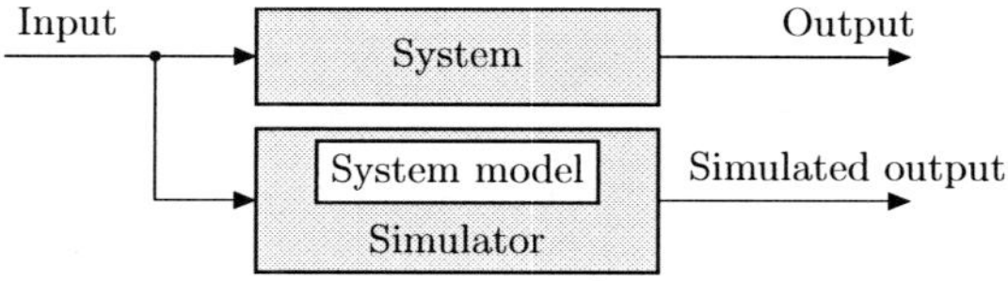

Figure 4.8: Basic concept of simulation

Autonomous automaton. The autonomous nondeterministic automaton introduced in Section 3.3.4 is used now without the index "i" for simplicity of notation. Thus, it is given as $\widehat{\mathcal{N}} = (\mathcal{N}_z, \widehat{\boldsymbol{L}}^n, \mathcal{Z}_0)$ with the characteristic function $\chi_{\widehat{\boldsymbol{L}}^n}(z(k+1), z(k))$. The possibility to move to a certain successor state is obtained recursively by applying

$$Poss(z_p(k+1) = z(k+1)) = \bigvee_{z=1}^{N} \chi_{\widehat{\boldsymbol{L}}^n}(z(k+1), z(k)) \cdot Poss(z_p(k) = z(k)). \qquad (4.23)$$

To allow for an easy implementation, this calculation can be carried out in a vectorial form with the possibility distribution[1]

$$\boldsymbol{poss}(z(k)) = \begin{pmatrix} Poss(z_p(k) = 1) \\ \vdots \\ Poss(z_p(k) = N) \end{pmatrix} \qquad (4.24)$$

and the matrix

$$\widehat{\boldsymbol{L}}^n_{sim} = \qquad (4.25)$$

[1] This notion has been adapted from the probability distribution of a stochastic process [84].

$$\begin{pmatrix} Poss(z_p(k+1)=1, z_p(k)=1) & \dots & Poss(z_p(k+1)=1, z_p(k)=N) \\ \vdots & \ddots & \vdots \\ Poss(z_p(k+1)=N, z_p(k)=1) & \dots & Poss(z_p(k+1)=N, z_p(k)=N) \end{pmatrix}.$$

The behavior of an autonomous nondeterministic automaton is given by the following algorithm:

Algorithm 1: Behavior of an autonomous nondeterministic automaton

---

Given: Initialized autonomous automaton $\widehat{\mathcal{N}} = (\mathcal{N}_z, \widehat{\boldsymbol{L}}^n, \mathcal{Z}_0)$.

Initialize

| $\boldsymbol{poss}(z(0)) = \mathbf{1}\ \forall z(0) \in \mathcal{Z}_0$.

End

For $k_e + +$

| Apply $\boldsymbol{poss}(z(k_e+1)) = \widehat{\boldsymbol{L}}^n_{sim} \cdot \boldsymbol{poss}(z(k_e))$.

End

Result: Possibility distribution $\boldsymbol{poss}(z(k_e+1))$ of the successor state for $k_e = 0, 1, \dots$.

I/O-automata. Algorithm 1 can be extended to cope with the nondeterministic I/O-automaton without coupling signals $\tilde{\mathcal{N}} = (\mathcal{N}_z, \mathcal{N}_v, \mathcal{N}_w, \tilde{\boldsymbol{L}}^n, \mathcal{Z}_0)$ defined in Section 3.3.3 as follows[2]:

$$Poss(z_p(k+1) = z(k+1), w_p(k) = w(k)) = \tag{4.26}$$
$$\bigvee_{z=1}^{N} \bigvee_{v=1}^{M} \chi_{\tilde{L}^n}(z(k+1), w(k), z(k), v(k)) \cdot Poss(z_p(k) = z(k), v_p(k) = v(k)).$$

It is assumed that the input is independent of the system state such that $Poss(z(k), v(k)) = Poss(z(k)) \cdot Poss(v(k))$ holds. This relation can be transformed into the possibility distribution $\boldsymbol{poss}(z(k), v(k)) = \boldsymbol{poss}(z(k)) \otimes \boldsymbol{poss}(v(k))$[3] by applying eqn. (4.24) to the input possibility. The matrix

$$\tilde{\boldsymbol{L}}^n_{sim} = \begin{pmatrix} \tilde{\boldsymbol{L}}(w_p(k)=1, v_p(k)=1) & \dots & \tilde{\boldsymbol{L}}(w_p(k)=1, v_p(k)=M) \\ \vdots & \ddots & \vdots \\ \tilde{\boldsymbol{L}}(w_p(k)=R, v_p(k)=1) & \dots & \tilde{\boldsymbol{L}}(w_p(k)=R, v_p(k)=M) \end{pmatrix} \tag{4.27}$$

consists of $M \cdot R$ blocks. Each block $\tilde{\boldsymbol{L}}(w_p(k) = i, v_p(k) = j)$ includes the possibilities of the state transitions given by eqn. (4.25) for the corresponding input/output pair. The behavior of a nondeterministic I/O-automaton without coupling signals is given by Algorithm 2.

[2] The index "i" is again omitted for simplicity of notation. In addition, faults are not taken into account because they are just another input signal of the I/O-automaton and do not bring up any further aspects in this context.

[3] $\otimes$ denotes the Hadamard product of two vectors which defines the element by element multiplication of two vectors.

Algorithm 2: Behavior of an I/O-automaton without coupling signals

Given: Initialized I/O-automaton $\tilde{\mathcal{N}} = (\mathcal{N}_z, \mathcal{N}_v, \mathcal{N}_w, \tilde{\boldsymbol{L}}^n, \mathcal{Z}_0)$,
Sequence of measured inputs $V(0 \ldots k_e)$.

Initialize
| $\boldsymbol{poss}(z(0)) = \mathbf{1}\ \forall z(0) \in \mathcal{Z}_0$.

End

For k_e ++
- Wait for $v(k_e)$.
- Get $\boldsymbol{poss}(v(k_e))$ from measurement.
- Apply $\boldsymbol{poss}(z(k_e), v(k_e)) = \boldsymbol{poss}(z(k_e)) \otimes \boldsymbol{poss}(v(k_e))$.
- Apply $\boldsymbol{poss}(z(k_e+1), w(k_e)) = \tilde{\boldsymbol{L}}^n_{sim} \cdot \boldsymbol{poss}(z(k_e), v(k_e))$.
- Apply $\boldsymbol{poss}(z(k_e+1)) = \bigvee\limits_{w(k_e)=1}^{R} \boldsymbol{poss}(z(k_e+1), w(k_e))$.
- Apply $\boldsymbol{poss}(w(k_e)) = \bigvee\limits_{z(k_e+1)=1}^{N} \boldsymbol{poss}(z(k_e+1), w(k_e))$.

End

Result: Possibility distribution $\boldsymbol{poss}(z(k_e+1))$ of the successor state and $\boldsymbol{poss}(w(k_e))$ of the output for $k_e = 0, 1, \ldots$.

4.3.2 Behavior of nondeterministic asynchronous networks of I/O-automata using online composition

Algorithm 2 can be extended directly to a nondeterministic asynchronous network of I/O-automata $\mathcal{NAN}$, if the equivalent I/O-automaton $\tilde{\mathcal{N}} = (\mathcal{N}_{\boldsymbol{z}}, \mathcal{N}_{\boldsymbol{v}}, \mathcal{N}_{\boldsymbol{w}}, \tilde{\boldsymbol{L}}^n, \mathcal{Z}_0)$ obtained by composition (Section 3.5.2) would be used. Even though this way sounds to be easy, it is not realizable because of the problem of state-space explosion. The straightforward solution to retain the memory advantage of the network description is to apply the composition rules defined in Section 3.5.2 to access and combine the relevant information for the current simulation step [78, 84]. This so-called online composition is gained by replacing the characteristic function $\chi_{\tilde{\boldsymbol{L}}^n}(\boldsymbol{z}(k+1), \boldsymbol{w}(k), \boldsymbol{z}(k), \boldsymbol{v}(k))$ in eqn. (4.26) with the composition rule (3.79) to obtain

$$
\begin{aligned}
&Poss(\boldsymbol{z}(k+1), \boldsymbol{w}(k)) \qquad (4.28)\\
&= \bigvee_{\boldsymbol{z}: Poss(\boldsymbol{z}(k))=1} \left(\bigvee_{\boldsymbol{r}=1}^{T} \bigwedge_{i=1}^{\nu} \chi_{\boldsymbol{L}_i^n}(z_i(k+1), \boldsymbol{w}_i(k), r_i(k), z_i(k), \boldsymbol{v}_i(k), \boldsymbol{k}_i^T \cdot \boldsymbol{r}(k)) \right)\\
&= \bigvee_{\boldsymbol{z}(k)} \left(\bigvee_{\boldsymbol{r}=1}^{T} \bigwedge_{i=1}^{\nu} \chi_{\boldsymbol{L}_i^n}(z_i(k+1), \boldsymbol{w}_i(k), r_i(k), z_i(k), \boldsymbol{v}_i(k), \boldsymbol{k}_i^T \cdot \boldsymbol{r}(k)) \right) \wedge Poss(\boldsymbol{z}(k))
\end{aligned}
$$

where the evaluation of $\bigvee\limits_{\boldsymbol{z}: Poss(\boldsymbol{z}(k))=1}$ is abbreviated by $\bigvee\limits_{\boldsymbol{z}(k)} \ldots \wedge Poss(\boldsymbol{z}(k))$ in the remainder of this thesis. The calculation of eqn. (4.28) is carried out for all combinations of $\boldsymbol{z}(k+1)$ and $\boldsymbol{w}(k)$ and inserted into the possibility distribution vector $\boldsymbol{poss}(\boldsymbol{z}(k+1), \boldsymbol{w}(k))$. The simulation with online composition is given in the following algorithm:

Algorithm 3: Behavior of a network of nondeterministic I/O-automata using online composition

---

Given: Initialized asynchronous network of I/O-automata $\mathcal{NAN}$,
Sequence of measured inputs $\boldsymbol{V}(0 \dots k_e)$.

Initialize
| $\boldsymbol{poss}(\boldsymbol{z}(0)) = \mathbf{1}\ \forall \boldsymbol{z}(0) \in \mathcal{Z}_0$.

End

For $k_e + +$
- Wait for $\boldsymbol{v}(k_e)$.
- Get $\boldsymbol{poss}(\boldsymbol{v}(k_e))$ from measurement.
- **For** *all combinations of* $\boldsymbol{z}(k_e+1)$ *and* $\boldsymbol{w}(k_e)$
  - Apply eqn. (4.28) to get $Poss(\boldsymbol{z}(k_e+1), \boldsymbol{w}(k_e))$.
  - Insert the result into the possibility distribution vector $\boldsymbol{poss}(\boldsymbol{z}(k_e+1), \boldsymbol{w}(k_e))$.
- **End**
- Apply $\boldsymbol{poss}(\boldsymbol{z}(k_e+1)) = \bigvee_{\boldsymbol{w}(k_e)=1}^{R} \boldsymbol{poss}(\boldsymbol{z}(k_e+1), \boldsymbol{w}(k_e))$.
- Apply $\boldsymbol{poss}(\boldsymbol{w}(k_e)) = \bigvee_{\boldsymbol{z}(k_e+1)=1}^{N} \boldsymbol{poss}(\boldsymbol{z}(k_e+1), \boldsymbol{w}(k_e))$.

End

Result: Possibility distribution $\boldsymbol{poss}(z(k_e+1))$ of the successor state and $\boldsymbol{poss}(\boldsymbol{w}(k_e))$ of the output for $k_e = 0, 1, \dots$.

Theorem 4.
The simulation Algorithm 3 using online composition returns at each time k_e the same possibility distribution $\boldsymbol{poss}(z(k_e+1))$ of the successor state and $\boldsymbol{poss}(\boldsymbol{w}(k_e))$ of the output of an nondeterministic asynchronous network of I/O-automata $\mathcal{NAN}$ as the simulation Algorithm 2 using the equivalent I/O-automaton $\tilde{\mathcal{N}}$ which has been obtained by off-line composition. □

Proof. The proof follows directly from the construction procedure of Algorithm 3 and Theorem 3. ■

4.4 Complexity considerations

Qualitative models as I/O-automata suffer from the problem of *state space explosion* [27, 33, 62, 70, 84, 120]. The high complexity of these models results from the necessity to explicitly store all possible transitions of the considered system, e.g. in tables, as there exists usually no closed expression for the characteristic function of the behavioral relation. The amount of data that needs to be stored for a monolithic description of a system increases exponentially with a linear increasing number of components such that the resulting model becomes quickly unmanageable even with modern computers. A solution to this problem is the usage of a structured model of the system as the asynchronous networks of I/O-automata defined in Chapter 3. The behavior of the system is then given in terms of the component behaviors and a coupling law. Hence, the amount of storage capacity is in general considerably reduced because the dependencies between the components must not be considered explicitly in the monolithic model but are implicitly given by the network topology. The following example is used to illustrate these complexity issues.

Example 7 *Complexity issues of discrete-event modeling*

First, a fairly simple network is considered. It consists of one component whose interconnection output is coupled with its interconnection input as shown in Fig. 4.9(a).

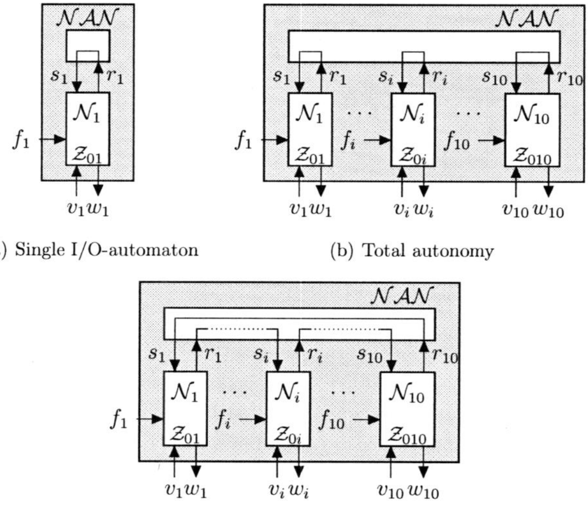

(a) Single I/O-automaton

(b) Total autonomy

(c) Total heteronomy

Figure 4.9: Different networks of I/O-automata

For simplicity, all signals of the I/O-automaton are assumed to have 10 *values, i.e.* $N_i = 10, M_i = 10, R_i = 10, P_i = 10, T_i = 10$ *and* $S_i = 10$. *The behavioral relation* $\boldsymbol{L}_i$ *is defined in the worst case for all values of each signal. Thus, the following number of possible transitions has to be stored:*

$$N_i^2 \cdot M_i \cdot R_i \cdot P_i \cdot T_i \cdot S_i = 10^7. \tag{4.29}$$

This number can be reduced by composing the network to a single I/O-automaton using eqn. (3.78) or (3.79) to

$$N_i^2 \cdot M_i \cdot R_i \cdot S_i = 10^5. \tag{4.30}$$

Thus, composition is useful in this setting. Next, a network of ten previously described components in total autonomy shown in Fig. 4.9(b) is investigated. The amount of data to be stored is given as the sum of the component data calculated in eqn. (4.29):

$$\sum_{i=1}^{10} N_i^2 \cdot M_i \cdot R_i \cdot P_i \cdot T_i \cdot S_i = 10^8. \tag{4.31}$$

By composing only the components, it can be reduced according to eqn. (4.30) to

$$\sum_{i=1}^{10} N_i^2 \cdot M_i \cdot R_i \cdot S_i = 10^6. \tag{4.32}$$

If the whole network is composed to one I/O-automaton by using the composition rules (3.78) or (3.79), the number of entries of the behavioral relation is given as:

$$\prod_{i=1}^{10} N_i^2 \cdot M_i \cdot R_i \cdot S_i = \prod_{i=1}^{10} 1 \cdot 10^5 = 10^{50}. \tag{4.33}$$

Finally, a network of ten components in total heteronomy depicted in Fig. 4.9(c) will be analyzed. It consumes the same amount of data for the network description given by (4.31) and for the single I/O-automaton resulting from composition considered in (4.33). The difference to the case of total autonomy relies in the intermediate step because there is no possibility to just compose the component models like in eqn. (4.32). Hence, the amount of data that needs to be stored for the network cannot not be further reduced.

The example clarifies that the amount of data needed to store a network of I/O-automata is in general significantly lower than that of the corresponding composed single I/O-automaton. The latter might be unmanageable even with modern computers. Thus, a component-oriented modeling formalism has distinct advantages in discrete-event modeling of complex systems compared to a monolithic approach. The amount of data can be further reduced by analyzing the system structure. It has been shown in eqns. (4.30) and (4.32) that composition is useful, only if the interconnection output of a component is connected to its interconnection input.

Chapter 5

Centralized diagnosis of nondeterministic processes

The centralized diagnosis of nondeterministic processes using multi-models is investigated in this chapter. It bases on the method for state observation explained in 5.1 for a single nondeterministic I/O-automaton. This method is extended in Section 5.2 to diagnosis. The application of the centralized diagnostic method to nondeterministic asynchronous networks of I/O-automata using online composition is investigated in Section 5.3. This chapter concludes in Section 5.4.

5.1 State observation of single nondeterministic I/O-automata

The basic task of diagnosis is the consistency check between the measurements and the system model, i.e. the search for a path starting at the set of initial states whose arcs are weighted with the measured sequences of inputs and outputs (Fig. 5.1). This search implies to look for a corresponding state sequence to obtain the ideal observation result $\mathcal{Z}^{\star}(k_e) \in \mathcal{N}_z$

$$\mathcal{Z}^{\star}(k_e) := \{z(k_e) | B(k_e) \in \mathcal{B}_{\tilde{\mathcal{N}}}, \exists z(k_e+1) \in \mathcal{N}_z, f \in \mathcal{N}_f : \chi_{\tilde{L}^n}(z(k+1), w(k), z(k), v(k), f) = 1\} \tag{5.1}$$

which contains exactly those states $z(k_e)$ for which the model $\tilde{\mathcal{N}}$ is consistent with the measurements $B(k_e) = (V(0 \dots k_e), W(0 \dots k_e))$ up to time k_e. This set may contain more than one element. Hence, its elements are called *state candidates.* Note that the elements of $\mathcal{Z}^{\star}(k_e)$ do not depend upon the fault f because the aim is to observe the state and not the state in conjunction with the fault. The latter problem is not considered in this thesis. The notions of completeness and soundness known from diagnosis can be applied accordingly.

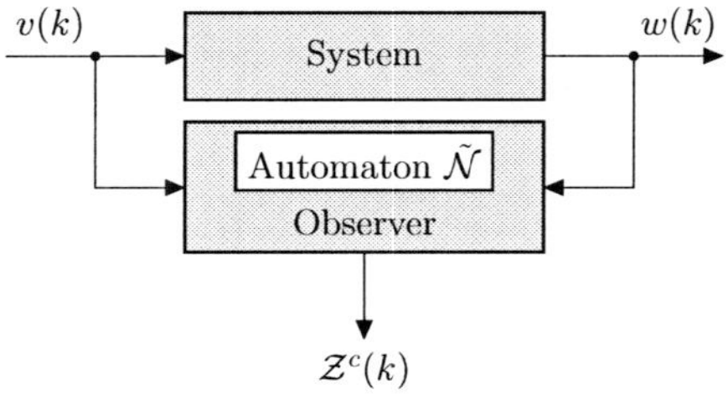

Figure 5.1: Centralized state observation of a discrete-event system

As the focus of this thesis is diagnosis of discrete-event systems, state observation will only be considered in the centralized setting for basic explanations. The terminology used here is motivated from continuous systems's theory and is not to be confused with the usual meaning in discrete-event systems theory where an observation denotes the measurement of a sensor and not the estimation of the systems state.

The approaches to centralized state observation and diagnosis in the next sections rely on [20, 70, 113] and on the notation introduced in [84]. They use the nondeterministic I/O-automaton (without coupling signals) $\tilde{\mathcal{N}} = (\mathcal{N}_z, \mathcal{N}_v, \mathcal{N}_w, \mathcal{N}_f, \tilde{\boldsymbol{L}}^n, \mathcal{Z}_0)$ where the index "i" is omitted for notational convenience and the measurements $B(k_e) = (V(0\dots k_e), W(0\dots k_e))$ for the consistency test to obtain the centrally obtained set of possible faults $\mathcal{Z}^c(k_e)$. The observation algorithm presented in the following is initialized with the a-priori initial condition $Poss(z(0), f) = 1\ \forall (z(0), f) \in \mathcal{Z}_0 \times \mathcal{N}_f$ that consists of all possible initial state and fault combinations. If the set of initial states is not given, $\mathcal{N}_z$ is used instead. Faults are assumed to be constant for notational convenience (Section 2.2). The centralized state observation problem is given for single nondeterministic I/O-automata as follows:

Centralized state observation problem for nondeterministic I/O-automata	
Given:	Initialized nondeterministic I/O-automaton $\tilde{\mathcal{N}} = (\mathcal{N}_z, \mathcal{N}_v, \mathcal{N}_w, \mathcal{N}_f, \tilde{\boldsymbol{L}}^n, \mathcal{Z}_0)$
	Measurements $B(k_e) = (V(0\dots k_e), W(0\dots k_e))$
Find:	Centrally obtained set of possible states $\mathcal{Z}^c(k_e)$

The solution to the centralized state observation problem is given recursively to ensure online applicability, i.e. the centrally obtained set of possible states is updated according to an increment of the discrete counter k_e. The calculation is divided into two parts. The *prediction step*

$$Poss(z(k+1), f, V(0\dots k), W(0\dots k)) \tag{5.2}$$
$$= \bigvee_{z(k):Poss(z(k),f,V(0\dots k-1),W(0\dots k-1))=1} \chi_{\tilde{\boldsymbol{L}}^n}(z(k+1), w(k), z(k), v(k), f).$$

evaluates recursively for each time k up to k_e the possibility of $z(k+1)$ being the successor state and f the fault based on the result of the previous prediction step given by all states $z(k)$ for which $Poss(z(k), f, V(0\ldots k-1), W(0\ldots k-1)) = 1$ holds, if the sequences of input values $V(0\ldots k)$ and output values $W(0\ldots k)$ have been measured. The initial condition $Poss(z(0), f) = 1\ \forall (z(0), f) \in \mathcal{Z}_0 \times \mathcal{N}_f$ is used in the first step ($k = 0$):

$$Poss(z(1), f, v(0), w(0)) = \bigvee_{z(0):Poss(z(0),f)=1} \chi_{\tilde{\boldsymbol{L}}^n}(z(1), w(0), z(0), v(0), f). \tag{5.3}$$

The *projection step*

$$\begin{aligned} &Poss(z(k), V(0\ldots k), W(0\ldots k)) \\ &= \bigvee_{z(k+1)\in\mathcal{N}_z} \bigvee_{f:Poss(z(k),f,V(0\ldots k-1),W(0\ldots k-1))=1} \chi_{\tilde{\boldsymbol{L}}^n}(z(k+1), w(k), z(k), v(k), f) \end{aligned} \tag{5.4}$$

returns the possible states $z(k)$ under the condition that the sequences of input values $V(0\ldots k)$ and output values $W(0\ldots k)$ have been measured. The centrally obtained observation result is given as the centrally obtained set of states that are possible at time k:

$$\mathcal{Z}^c(k) = \{z(k) | Poss(z(k), V(0\ldots k), W(0\ldots k)) = 1\}. \tag{5.5}$$

The centralized observation algorithm for a single I/O-automaton is given by the following algorithm:

Algorithm 4: Centralized state observation of a single nondeterministic I/O-automaton

---

Given: Initialized nondeterministic I/O-automaton $\tilde{\mathcal{N}} = (\mathcal{N}_z, \mathcal{N}_v, \mathcal{N}_w, \mathcal{N}_f, \tilde{\boldsymbol{L}}^n, \mathcal{Z}_0)$,
Measurements $B(k_e) = (V(0\ldots k_e), W(0\ldots k_e))$.

Initialize
| $Poss(z(0), f) = 1\ \forall (z(0), f) \in \mathcal{Z}_0 \times \mathcal{N}_f$.
End
For $k_e + +$
| Wait for $(v(k_e)/w(k_e))$.
| **For** *all combinations of* $z(k_e+1)$ *and* f
| | Do prediction by applying eqns. (5.2) and (5.3).
| **End**
| **For** *all combinations of* $z(k_e)$
| | Do projection by applying eqn. (5.4).
| **End**
| Apply eqn. (5.5) to obtain $\mathcal{Z}^c(k_e)$.
End
Result: Centrally obtained set of possible states $\mathcal{Z}^c(k_e)$ for $k_e = 0, 1, \ldots$.

Theorem 5.
The observation result $\mathcal{Z}^c(k_e)$ obtained by Algorithm 4 is complete and sound

$$\mathcal{Z}^c(k_e) = \mathcal{Z}^\star(k_e) \tag{5.6}$$

for $k_e = 0, 1, \ldots$. □

Proof. The proof of Theorem 5 can be held in analogy to the proof of Theorem 6 and is therefore omitted. ■

The centrally obtained observation result cannot be further improved based on the given measurements and model as it is equal to the ideal observation result. Thus, neither spurious solutions are included in $\mathcal{Z}^c(k_e)$ nor are states missing. The only way to achieve a better result is given by adding measurements or enhancing the model but not by altering the observation algorithm.

5.2 Diagnosis of single nondeterministic I/O-automata

The extension of the observation method for a single I/O-automaton to diagnosis is given in this section. It uses the centralized information structure shown in Fig. 5.2 which is the least restrictive set-up for diagnosis. A single diagnostic unit D^c applies the principle of consistency-based diagnosis to the model of the entire system and the measurements of all inputs and outputs. The faults whose models are consistent with the measurements at each time k are included in the centrally obtained diagnostic result $\mathcal{F}^c(k)$. This result is preferred to be as close to the ideal diagnostic result $\mathcal{F}^\star(k)$ as possible but needs to be complete, i.e. no fault should be excluded wrongly.

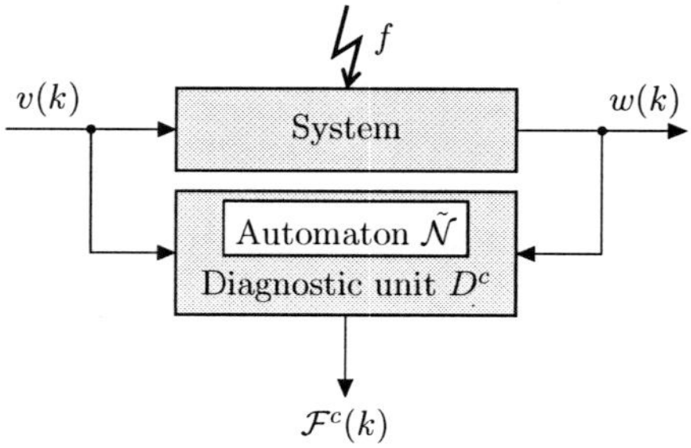

Figure 5.2: Centralized diagnosis of a discrete-event system

The diagnostic algorithm is initialized, like the observation algorithm, with the a-priori initial condition $Poss(z(0), f) = 1 \; \forall (z(0), f) \in \mathcal{Z}_0 \times \mathcal{N}_f$ that consists of all possible initial state and fault combinations. If the set of initial states is not given, $\mathcal{N}_z$ is used instead. Faults are assumed to be constant for notational convenience (Section 2.2). The centralized fault diagnostic problem can be stated as follows:

Centralized fault diagnostic problem for nondeterministic I/O-automata	
Given:	Initialized nondeterministic I/O-automaton $\tilde{\mathcal{N}} = (\mathcal{N}_z, \mathcal{N}_v, \mathcal{N}_w, \mathcal{N}_f, \tilde{\boldsymbol{L}}^n, \mathcal{Z}_0)$
	Measurements $B(k_e) = (V(0\ldots k_e), W(0\ldots k_e))$
Find:	Centrally obtained set of possible faults $\mathcal{F}^c(k_e)$

The solution is split up into two parts as known from observation. The prediction step (5.2) is given without changes recursively as

$$\begin{aligned} &Poss(z(k+1), f, V(0\ldots k), W(0\ldots k)) \\ &= \bigvee_{z(k):Poss(z(k),f,V(0\ldots k-1),W(0\ldots k-1))=1} \chi_{\tilde{\boldsymbol{L}}^n}(z(k+1), w(k), z(k), v(k), f) \end{aligned} \tag{5.7}$$

based on the result of the previous prediction step given by all states $z(k)$ for which $Poss(z(k), f, V(0\ldots k-1), W(0\ldots k-1)) = 1$ holds. It is initialized for $k = 0$ according to eqn. (5.3) as:

$$Poss(z(1), f, v(0), w(0)) = \bigvee_{z(0):Poss(z(0),f)=1} \chi_{\tilde{\boldsymbol{L}}^n}(z(1), w(0), z(0), v(0), f). \tag{5.8}$$

This result is used in the *projection step*

$$\begin{aligned} &Poss(f, V(0\ldots k), W(0\ldots k)) \\ &= \bigvee_{z(k+1)\in\mathcal{N}_z} \bigvee_{z(k):Poss(z(k),f,V(0\ldots k-1),W(0\ldots k-1))=1} \chi_{\tilde{\boldsymbol{L}}^n}(z(k+1), w(k), z(k), v(k), f) \\ &= \bigvee_{z(k+1)\in\mathcal{N}_z} Poss(z(k+1), f, V(0\ldots k), W(0\ldots k)) \end{aligned} \tag{5.9}$$

to obtain all possible faults f under the condition that the sequences of input values $V(0\ldots k)$ and output values $W(0\ldots k)$ have been measured. The diagnostic result is given as the centrally obtained set of faults that are possible at time k:

$$\mathcal{F}^c(k) = \{f | Poss(f, V(0\ldots k), W(0\ldots k)) = 1\}. \tag{5.10}$$

The centralized diagnostic algorithm for a single I/O-automaton is given by Algorithm 5.

Remark. *Adding a test for the "total inconsistency" after the prediction step has been proven to be useful for the implementation of Algorithm 5 [84]:*

If $Poss(z(k_e+1), f, V(0\ldots k_e), W(0\ldots k_e)) = 0,\ \forall z(k_e+1),\ f$
restart with initial condition and notify the operator.

This condition indicates that the measurement $B(k_e)$ is inconsistent with the modeled system behavior of all faults. Thus, the closed world assumption demanded in Section 2.1

Algorithm 5: Centralized diagnosis of a single nondeterministic I/O-automaton

Given: Initialized nondeterministic I/O-automaton $\tilde{\mathcal{N}} = (\mathcal{N}_z, \mathcal{N}_v, \mathcal{N}_w, \mathcal{N}_f, \tilde{\boldsymbol{L}}^n, \mathcal{Z}_0)$,
Measurements $B(k_e) = (V(0 \dots k_e), W(0 \dots k_e))$.

Initialize
| $Poss(z(0), f) = 1 \; \forall (z(0), f) \in \mathcal{Z}_0 \times \mathcal{N}_f$.
End
For k_e++
| Wait for $(v(k_e)/w(k_e))$.
| **For** *all combinations of* f
| | **For** *all combinations of* $z(k_e+1)$
| | | Do prediction by applying eqns. (5.7) and (5.8).
| | **End**
| | Do projection by applying eqn. (5.9).
| **End**
| Apply eqn. (5.10) to obtain $\mathcal{F}^c(k_e)$.
End
Result: Centrally obtained set of possible faults $\mathcal{F}^c(k_e)$ for $k_e = 0, 1, \dots$.

is violated either due to the occurrence of an unconsidered fault $f \notin \mathcal{N}_f$*, an uncomplete model such that relation* (2.13) *does no longer hold or a single mismeasurement. The result is always an empty centrally obtained set of possible faults* $\mathcal{F}^c(k_e) = \emptyset$ *for all further recursions. The algorithm is restarted and the operator is informed to ensure actionability of diagnosis.* □

Theorem 6.
The diagnostic result $\mathcal{F}^c(k_e)$ *obtained by Algorithm 5 is complete and sound*

$$\mathcal{F}^c(k_e) = \mathcal{F}^\star(k_e) \tag{5.11}$$

for $k_e = 0, 1, \dots$. □

Proof. Equation (5.11) can be inserted into the definition of consistency (2.7) to obtain

$$f \in \mathcal{F}^c(k_e) \Leftrightarrow B(k_e) \in \mathcal{B}_{\tilde{\mathcal{N}}}(f). \tag{5.12}$$

It has to be shown that this relation holds for each $0 \le k \le k_e$ to prove the theorem.

The right-hand side of eqn. (5.12) implies that there exists for the fault f a state sequence $Z(0 \dots k_e+1)$ given by eqn. (3.25) such that $\chi_{\tilde{\boldsymbol{L}}^n}(z(k+1), w(k), z(k), v(k), f) = 1$ holds for each k with $z(0) \in \mathcal{Z}_0$. This statement is equivalent to the relation

$$\exists Z(0 \dots k_e+1) : \bigwedge_{k=0}^{k_e} \chi_{\tilde{\boldsymbol{L}}^n}(z(k+1), w(k), z(k), v(k), f) = 1 \tag{5.13}$$

where $z(k)$ are the elements of $Z(0 \dots k_e + 1)$. Such a state sequence exists for the fault f, if the following condition is satisfied:

$$\exists z(0) \in \mathcal{Z}_0 : \bigvee_{z(k_e+1)} \bigvee_{z(k_e)} \cdots \bigvee_{z(0)} \bigwedge_{k=0}^{k_e} \chi_{\tilde{L}^n}(z(k+1), w(k), z(k), v(k), f) = 1. \tag{5.14}$$

The information about the initial state and the considered fault is given by $Poss(z(0), f)$ such that eqn. (5.14) can be rewritten as

$$\bigvee_{z(k_e+1)} \bigvee_{z(k_e)} \cdots \bigvee_{z(0):Poss(z(0),f)=1} \bigwedge_{k=0}^{k_e} \chi_{\tilde{L}^n}(z(k+1), w(k), z(k), v(k), f) = 1. \tag{5.15}$$

The condition on the left-hand side of eqn. (5.12) can be rewritten using eqn. (5.9) to obtain the relation

$$f \in \mathcal{F}^c(k_e) \Leftrightarrow Poss(f, V(0 \dots k_e), W(0 \dots k_e)) = 1. \tag{5.16}$$

Thus, to prove the theorem, the equality

$$\begin{aligned} &Poss(f, V(0 \dots k_e), W(0 \dots k_e)) \\ &= \bigvee_{z(k_e+1)} \bigvee_{z(k_e)} \cdots \bigvee_{z(0):Poss(z(0),f)=1} \bigwedge_{k=0}^{k_e} \chi_{\tilde{L}^n}(z(k+1), w(k), z(k), v(k), f) \end{aligned} \tag{5.17}$$

has to be proven for each $0 \leq k \leq k_e$. This proof is held by induction.

Induction basis. For $k_e = 0$, inserting eqn. (5.8) into eqn. (5.9) yields

$$\begin{aligned} Poss(f, v(0)), w(0)) &= \bigvee_{z(1)} Poss(z(1), f, v(0), w(0)) \\ &= \bigvee_{z(1)} \bigvee_{z(0):Poss(z(0),f)=1} \chi_{\tilde{L}^n}(z(1), w(0), z(0), v(0), f) \end{aligned}$$

which is identical to the right-hand side of eqn. (5.17) for $k_e = 0$ and proves the induction basis.

Inductive step. The inductive hypothesis is proposed by assuming that equation (5.17) holds for some $k_e = k_h$

$$\begin{aligned} &Poss(f, V(0 \dots k_h), W(0 \dots k_h)) \\ &= \bigvee_{z(k_h+1)} \bigvee_{z(k_h)} \cdots \bigvee_{z(0):Poss(z(0),f)=1} \bigwedge_{k=0}^{k_h} \chi_{\tilde{L}^n}(z(k+1), w(k), z(k), v(k), f). \end{aligned} \tag{5.18}$$

It has to be proven that eqn. (5.17) is satisfied for $k_e = k_h + 1$ based on this hypothesis

$$\begin{aligned} &Poss(f, V(0 \dots k_h + 1), W(0 \dots k_h + 1)) \\ &= \bigvee_{z(k_h+2)} \bigvee_{z(k_h+1)} \cdots \bigvee_{z(0):Poss(z(0),f)=1} \bigwedge_{k=0}^{k_h+1} \chi_{\tilde{L}^n}(z(k+1), w(k), z(k), v(k), f). \end{aligned} \tag{5.19}$$

The right-hand side of this equation can be reformulated as

$$
\begin{aligned}
&\bigvee_{z(k_h+2)} \bigvee_{z(k_h+1)} \bigvee_{z(k_h)} \cdots \bigvee_{z(0):Poss(z(0),f)=1} \bigwedge_{k=0}^{k_h+1} \chi_{\tilde{L}^n}(z(k+1), w(k), z(k), v(k), f) \\
&= \bigvee_{z(k_h+2)} \chi_{\tilde{L}^n}(z(k_h+2), w(k_h+1), z(k_h+1), v(k_h+1), f) \\
&\wedge \left(\bigvee_{z(k_h+1)} \bigvee_{z(k_h)} \cdots \bigvee_{z(0):Poss(z(0),f)=1} \bigwedge_{k=0}^{k_h} \chi_{\tilde{L}^n}(z(k+1), w(k), z(k), v(k), f) \right) \\
&= \bigvee_{z(k_h+2)} \chi_{\tilde{L}^n}(z(k_h+2), w(k_h+1), z(k_h+1), v(k_h+1), f) \\
&\wedge Poss(f, V(0 \dots k_h), W(0 \dots k_h)) \\
&= Poss(f, V(0 \dots k_h+1), W(0 \dots k_h+1))
\end{aligned}
$$

which proves the inductive step and the theorem. ■

The centrally obtained diagnostic result cannot be further improved like the centrally obtained observation result. Neither spurious solutions are included in $\mathcal{F}^c(k_e)$ nor are faults missing. The only way to achieve a better result is again given by adding measurements or enhancing the model but not by altering the diagnostic algorithm.

5.3 Application of the centralized diagnostic algorithm to nondeterministic asynchronous networks of I/O-automata using online composition

The diagnosis of an nondeterministic asynchronous network of I/O-automata $\mathcal{NAN}$ is investigated in this section using the centralized information structure. The single diagnostic unit D^c uses a monolithic model of the system and the measurements of all inputs and outputs $\boldsymbol{B}(k_e) = (\boldsymbol{V}(0 \dots k_e), \boldsymbol{W}(0 \dots k_e))$ to obtain the centrally obtained diagnostic result $\mathcal{F}^c(k_e)$ which is again preferred to be as close to the ideal diagnostic result as possible (Fig. 5.3). The initial conditions of the network are given for all ν I/O-automata $\mathcal{N}_i$ separately as $Poss(z_i(0), f_i) = 1 \ \forall (z_i(0), f_i) \in \mathcal{Z}_{0i} \times \mathcal{N}_{f_i}$. The centralized fault diagnostic problem for networks of I/O-automata can be stated as follows:

	Centralized fault diagnostic problem for nondeterministic asynchronous networks of I/O-automata
Given:	Initialized nondeterministic asynchronous network of I/O-automata $\mathcal{NAN}$
	Measurements $\boldsymbol{B}(k_e) = (\boldsymbol{V}(0 \dots k_e), \boldsymbol{W}(0 \dots k_e))$
Find:	Centrally obtained set of possible faults $\mathcal{F}^c(k_e)$

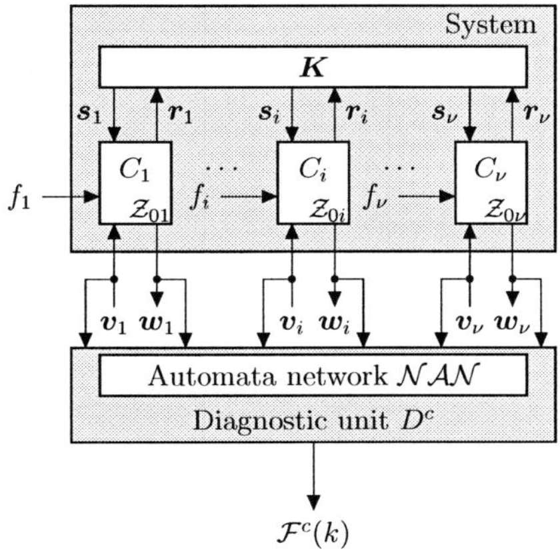

Figure 5.3: Centralized diagnosis of an interconnected discrete-event system

The straightforward solution to the centralized fault diagnostic problem for nondeterministic networks of I/O-automata would be to use the equivalent I/O-automaton $\tilde{\mathcal{N}} = (\mathcal{N}_{\boldsymbol{z}}, \mathcal{N}_{\boldsymbol{v}}, \mathcal{N}_{\boldsymbol{w}}, \tilde{\boldsymbol{L}}^n, \mathcal{Z}_0)$ obtained by (off-line) composition with an extension of Algorithm 5 to vectorial signals. Even though this way sounds to be easy, it is not realizable because of the problem of state-space explosion. The only way to retain the memory advantage of the network description is to apply the composition rule online to access and combine the relevant information for the current diagnostic step [77, 78, 84].

The online composition has already been used in Section 4.3.2 for the simulation of the network. The characteristic function $\chi_{\tilde{\boldsymbol{L}}^n}(\boldsymbol{z}(k+1), \boldsymbol{w}(k), \boldsymbol{z}(k), \boldsymbol{v}(k))$ is replaced in eqns. (5.7) and (5.8) with the composition rule (3.79) to obtain the new prediction step

$$Poss(\boldsymbol{z}(k+1), \boldsymbol{f}, \boldsymbol{V}(0\ldots k), \boldsymbol{W}(0\ldots k)) \tag{5.20}$$
$$= \bigvee_{\boldsymbol{z}(k):Poss(\boldsymbol{z}(k),\boldsymbol{f},\boldsymbol{V}(0\ldots k-1),\boldsymbol{W}(0\ldots k-1))=1} \left(\bigvee_{\boldsymbol{r}(k)=1}^{T} \bigwedge_{i=1}^{\nu} \chi_{\boldsymbol{L}_i^n}(z_i(k+1), \boldsymbol{w}_i(k), \boldsymbol{r}_i(k), z_i(k), \boldsymbol{v}_i(k), \boldsymbol{k}_i^T \cdot \boldsymbol{r}(k), f_i) \right)$$

which is based on the result of the previous prediction step given by all states $\boldsymbol{z}(k)$ for which $Poss(\boldsymbol{z}(k), \boldsymbol{f}, \boldsymbol{V}(0\ldots k-1), \boldsymbol{W}(0\ldots k-1)) = 1$ holds. The new prediction step is initialized in the first step ($k_e = 0$) as

$$Poss(\boldsymbol{z}(1), \boldsymbol{f}, \boldsymbol{v}(0), \boldsymbol{w}(0)) \tag{5.21}$$
$$= \bigvee_{\boldsymbol{z}(0):\bigwedge_{i=1}^{\nu} Poss(z_i(0),f_i)} \left(\bigvee_{\boldsymbol{r}(0)=1}^{T} \bigwedge_{i=1}^{\nu} \chi_{\boldsymbol{L}_i^n}(z_i(1), \boldsymbol{w}_i(0), \boldsymbol{r}_i(0), z_i(0), \boldsymbol{v}_i(0), \boldsymbol{k}_i^T \cdot \boldsymbol{r}(0), f_i) \right)$$

where the relation

$$Poss(\boldsymbol{z}(0), \boldsymbol{f}) = \bigwedge_{i=1}^{\nu} Poss(z_i(0), f_i) \tag{5.22}$$

has been used. The extension of (5.9) to vectorial signals yields the projection step

$$\begin{aligned} &Poss(\boldsymbol{f}, \boldsymbol{V}(0 \ldots k), \boldsymbol{W}(0 \ldots k)) \\ &= \bigvee_{\boldsymbol{z}(k+1)} Poss(\boldsymbol{z}(k+1), \boldsymbol{f}, \boldsymbol{V}(0 \ldots k), \boldsymbol{W}(0 \ldots k)). \end{aligned} \tag{5.23}$$

The centrally obtained diagnostic result is given as the centrally obtained set of possible faults

$$\mathcal{F}^c(k) = \{\boldsymbol{f} | Poss(\boldsymbol{f}, \boldsymbol{V}(0 \ldots k), \boldsymbol{W}(0 \ldots k)) = 1\}. \tag{5.24}$$

This procedure results in the following algorithm:

Algorithm 6: Centralized diagnosis of a nondeterministic asynchronous network of I/O-automata using online composition

Given: Initialized nondeterministic asynchronous network of I/O-automata $\mathcal{NAN}$ given by eqn. (3.72),
Measurements $\boldsymbol{B}(k_e) = (\boldsymbol{V}(0 \ldots k_e), \boldsymbol{W}(0 \ldots k_e))$.

Initialize
| $Poss(z_i(0), f_i) = 1 \ \forall (z_i(0), f_i) \in \mathcal{Z}_{0i} \times \mathcal{N}_{f_i}$.
End
For $k_e{+}{+}$
  Wait for $(\boldsymbol{v}(k_e)/\boldsymbol{w}(k_e))$.
  For *all combinations of* $\boldsymbol{f}$
    For *all combinations of* $\boldsymbol{z}(k_e+1)$
      Do prediction with online composition by applying eqns. (5.20) and (5.21).
    End
    Do projection by applying eqn. (5.23).
  End
  Apply eqn. (5.24) to obtain $\mathcal{F}^c(k_e)$.
End

Result: Centrally obtained set of possible faults $\mathcal{F}^c(k_e)$ for $k_e = 0, 1, \ldots$.

Theorem 7.
The diagnostic result $\mathcal{F}^c(k_e)$ obtained by Algorithm 6 is complete and sound

$$\mathcal{F}^c(k_e) = \mathcal{F}^{\star}(k_e) \tag{5.25}$$

for $k_e = 0, 1, \ldots$. □

Proof. Relation (5.12) holds for each fault for the single I/O-automaton $\tilde{\mathcal{N}}$. As the behavior of the network $\mathcal{B}_{\mathcal{NAN}}(\boldsymbol{f})$ and of the single I/O-automaton $\mathcal{B}_{\tilde{\mathcal{N}}}(\boldsymbol{f})$ obtained by composition are equal for each fault $\boldsymbol{f}$ (Theorem 3), eqn. (5.12) has to hold for $\mathcal{B}_{\mathcal{NAN}}(\boldsymbol{f})$:

$$\boldsymbol{f} \in \mathcal{F}^c(k_e) \Leftrightarrow \boldsymbol{B}(k_e) \in \mathcal{B}_{\mathcal{NAN}}(\boldsymbol{f}). \tag{5.26}$$

Hence, the diagnostic result $\mathcal{F}^c(k_e)$ obtained by Algorithm 6 is proven to be complete and sound because it is equal to the diagnostic result given by Algorithm 5 which is complete and sound (Theorem 6). ■

Centralized diagnosis has been investigated in this section. It has been shown that the ideal diagnostic result can be reached using Algorithm 5 with the drawback that it is not directly applicable to the equivalent I/O-automaton of a network of I/O-automata because this I/O-automaton is in general not computable. Online composition of the relevant part of the characteristic function of the behavioral relation has been used to overcome this problem. Thus, the resulting Algorithm 6 can be applied to diagnosis of networks of I/O-automata. It should be mentioned at this point, that the problem of complexity reduction has not been solved yet but only relocated from a high memory consumption to a high number of mathematical operations because no structural information has been used in these algorithms.

5.4 Evaluation of results

The application of the principle of consistency-based diagnosis to discrete-event systems has been shown in this chapter based on a centralized information structure. The basic explanations have been given for a single I/O-automaton without coupling signals. It has been shown that the resulting algorithm can not be extended directly to diagnosis of nondeterministic asynchronous networks of I/O-automata based on the equivalent I/O-automaton because of the problem of state-space explosion. This problem has been solved by online composition of the relevant part of the characteristic function to retain computability. As composition accompanies the loss of network effects like state-dependent autonomy and asynchronous motion of several components, the presented algorithms do not differ from the diagnostic algorithms for synchronous networks of I/O-automata considered in [84].

Both algorithms have been proven to yield the *ideal diagnostic result* $\mathcal{F}^\star(k_e)$. The centrally obtained diagnostic result cannot be further improved by altering the diagnostic algorithm. They only way to enhance the diagnostic result is to design a more detailed model or to place additional senors. Despite the best achievable performance, both algorithms suffer from the problem of *scalability* because the number of calculations increases exponentially with the size of the single I/O-automaton or the network of I/O-automata. Altering the plant by adding or replacing components or changing the interconnection structure results in the need to re-build the model, if the first algorithm is used whereas only the network must be adapted for the second algorithm. Thus, the second algorithm has a higher *reusability* than the first one. In addition, the *reliability* of both approaches is low because diagnosis fails at all, if the centralized diagnostic unit fails. These considerations necessitate the need to design alternative methods for the diagnosis of large discrete-event systems.

Chapter 6

Decentralized diagnosis of nondeterministic asynchronous networks of I/O-automata

The decentralization of the diagnostic task is motivated in Section 6.1. The decentralized diagnostic algorithm is derived in Section 6.2. It is analyzed in Section 6.3 w.r.t simplifications resulting from autonomy for the general case allowing for interaction between the components. The complexity of the diagnostic algorithm is investigated in Section 6.4. This chapter concludes in Section 6.5.

6.1 Motivation for decentralized diagnosis

The diagnostic task of the overall system is split up into several smaller tasks in decentralized diagnosis. Each component of the system is associated with an own local diagnostic unit D_i^{dc} which has a model of the component in terms of a nondeterministic I/O-automaton $\mathcal{N}_i$. Each local diagnostic unit has only access to the control inputs $\boldsymbol{v}_i(k)$ and outputs $\boldsymbol{w}_i(k)$ of the corresponding component. The interconnection inputs $\boldsymbol{s}_i(k)$ and outputs $\boldsymbol{r}_i(k)$ of the component are assumed to be immeasurable. Thus, the measurements of the component are given according to eqn. (3.10) as $\tilde{\boldsymbol{B}}_i(k_e) = (\boldsymbol{V}_i(0 \dots k_e), \boldsymbol{W}_i(0 \dots k_e))$. All locally obtained diagnostic results $\mathcal{F}_i^{dc}(k)$ are combined for $0 \leq k \leq k_e$ in the so-called merger to obtain the decentrally obtained diagnostic result $\mathcal{F}^{dc}(k)$.

The decentralized information structure is depicted in Fig. 6.1. It is the most restrictive approach to diagnosis because the lowest amount of information of the system is available at each local diagnostic unit. This set-up is good scalable and reliable. It is expected that the decentrally obtained diagnostic result is identical to the centrally obtained diagnostic result in those states where the components work autonomously. A degradation of the diagnostic result is expected in the case of heteronomy, i.e. interaction between components, due to the disregarded interconnection signals. Nevertheless, the decentrally obtained diagnostic

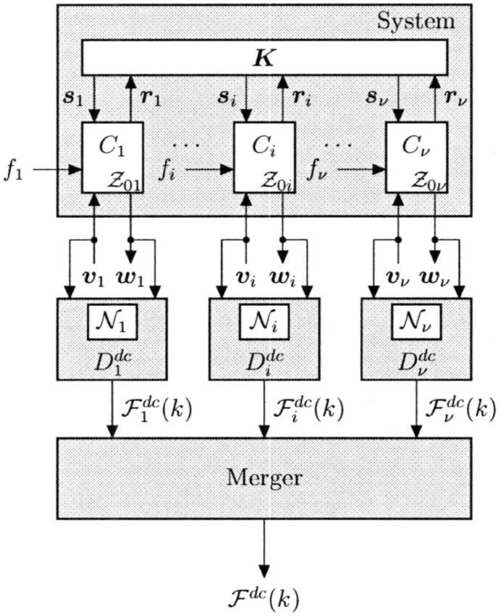

Figure 6.1: Decentralized diagnosis of an interconnected discrete-event system

result is demanded to be complete such that the relation $\mathcal{F}^{dc}(k) \supseteq \mathcal{F}^{\star}(k)$ must hold at each time step k. The aim of this chapter is to develop algorithms that yield the less possible amount of spurious solutions $\boldsymbol{f} \in \mathcal{F}^{spur}(k)$ given by eqn. (2.12).

6.2 Solution of the decentralized diagnostic problem

The decentralized diagnostic problem can be divided into two parts. First, the separat diagnosis of a component of the network using the local diagnostic unit D_i^{dc} is investigated. The decentralized diagnosis of all ν components of the system yields ν locally obtained diagnostic results $\mathcal{F}_1^{dc}(k), \ldots, \mathcal{F}_\nu^{dc}(k)$. The second part deals with the combination of these result to the diagnostic result of the whole system in the merger.

Local diagnosis. The problem of the local diagnosis of the components without regarding the interaction with other components can be stated as follows:

Local fault diagnostic problem for nondeterministic I/O-automata	
Given:	Initialized nondeterministic I/O-automaton $\mathcal{N}_i$
	Local measurements $\tilde{\boldsymbol{B}}_i(k_e) = (\boldsymbol{V}_i(0\ldots k_e), \boldsymbol{W}_i(0\ldots k_e))$
Find:	Locally obtained set of possible faults $\mathcal{F}_i^{dc}(k_e)$

The main task in solving the local fault diagnostic problem is to handle the immeasurable interconnection signals correctly. As they are unknown to the local diagnostic unit, the influence of the remaining network on this component cannot be taken into account for diagnostic purposes. It has been stated in [84] that the only way to handle these uncertainties while assuring a complete diagnostic result is to *consider all possible values* of the interconnection signals. Randomly choosing some values or taking the most probable one results in general in an incomplete diagnostic result. Thus, the aim of local diagnosis is to obtain the ideal locally obtained set of fault candidates

$$\mathcal{F}_i^{\star}(k_e) := \{f_i | \tilde{\boldsymbol{B}}_i(k_e) \in \tilde{\mathcal{B}}_{\mathcal{N}_i}(f_i)\} \tag{6.1}$$

that contains exactly those faults f_i for which the corresponding I/O-automaton is consistent with the local measurements $\tilde{\boldsymbol{B}}_i(k_e) = (\boldsymbol{V}_i(0 \dots k_e), \boldsymbol{W}_i(0 \dots k_e))$ at time k_e.

This aim is achieved by altering Algorithm 5 to consider all values of the interconnection input $\boldsymbol{s}_i(k)$ and output $\boldsymbol{r}_i(k)$ in the prediction and projection step and combine the results via disjunctions. The prediction step

$$\begin{aligned} &Poss(z_i(k+1), f_i, \boldsymbol{V}_i(0 \dots k), \boldsymbol{W}_i(0 \dots k)) \\ &= \bigvee_{z_i(k): Poss(z_i(k), f_i, \boldsymbol{V}_i(0 \dots k-1), \boldsymbol{W}_i(0 \dots k-1))=1} \\ &\quad \bigvee_{\boldsymbol{r}_i(k) \in \mathcal{N}_{\boldsymbol{r}_i}} \bigvee_{\boldsymbol{s}_i(k) \in \mathcal{N}_{\boldsymbol{s}_i}} \chi_{\boldsymbol{L}_i^n}(z_i(k+1), \boldsymbol{w}_i(k), \boldsymbol{r}_i(k), z_i(k), \boldsymbol{v}_i(k), \boldsymbol{s}_i(k), f_i) \end{aligned} \tag{6.2}$$

evaluates recursively for each time k up to k_e the possibility of $z_i(k+1)$ being the successor state and f_i the fault based on the result of the previous prediction step given by all states $z_i(k)$ for which $Poss(z_i(k), f_i, \boldsymbol{V}_i(0 \dots k-1), \boldsymbol{W}_i(0 \dots k-1)) = 1$ holds, if the sequences of input values $\boldsymbol{V}_i(0 \dots k)$ and output values $\boldsymbol{W}_i(0 \dots k)$ have been measured. The initial condition $Poss(z_i(0), f_i) = 1 \ \forall \in (z_i(0), f_i) \in \mathcal{Z}_{0i} \times \mathcal{N}_{f_i}$ is used in the first step ($k = 0$):

$$\begin{aligned} &Poss(z_i(1), f_i, \boldsymbol{v}_i(0), \boldsymbol{w}_i(0)) \\ &= \bigvee_{z_i(0): Poss(z_i(0), f_i)=1} \bigvee_{\boldsymbol{r}_i(0) \in \mathcal{N}_{\boldsymbol{r}_i}} \bigvee_{\boldsymbol{s}_i(0) \in \mathcal{N}_{\boldsymbol{s}_i}} \chi_{\boldsymbol{L}_i^n}(z_i(1), \boldsymbol{w}_i(0), \boldsymbol{r}_i(0), z_i(0), \boldsymbol{v}_i(0), \boldsymbol{s}_i(0), f_i). \end{aligned} \tag{6.3}$$

This result is used in the *projection step*

$$\begin{aligned} &Poss(f_i, \boldsymbol{V}_i(0 \dots k), \boldsymbol{W}_i(0 \dots k)) \\ &= \bigvee_{z_i(k+1) \in \mathcal{N}_{z_i}} \bigvee_{z_i(k): \bigvee_{z_i(k+1)} Poss(z_i(k+1), f_i, \boldsymbol{V}_i(0 \dots k), \boldsymbol{W}_i(0 \dots k))} \\ &\quad \bigvee_{\boldsymbol{r}_i(k) \in \mathcal{N}_{\boldsymbol{r}_i}} \bigvee_{\boldsymbol{s}_i(k) \in \mathcal{N}_{\boldsymbol{s}_i}} \chi_{\boldsymbol{L}_i^n}(z_i(k+1), \boldsymbol{w}_i(k), \boldsymbol{r}_i(k), z_i(k), \boldsymbol{v}_i(k), \boldsymbol{s}_i(k), f_i) \end{aligned} \tag{6.4}$$

to obtain the possible faults f_i under the condition that the sequences of input values $\boldsymbol{V}_i(0 \dots k)$ and output values $\boldsymbol{W}_i(0 \dots k)$ have been measured. The locally obtained diagnostic result is given as the locally obtained set of faults that are possible at time k:

$$\mathcal{F}_i^{dc}(k) = \{f_i | Poss(f_i, \boldsymbol{V}_i(0 \dots k), \boldsymbol{W}_i(0 \dots k)) = 1\}. \tag{6.5}$$

The local diagnostic algorithm for an I/O-automaton $\mathcal{N}_i$ is given as follows:

Algorithm 7: Local diagnosis of a nondeterministic I/O-automaton

Given: Initialized nondeterministic I/O-automaton $\mathcal{N}_i$ given by eqn. (3.29),
Local measurements $\tilde{\boldsymbol{B}}_i(k_e) = (\boldsymbol{V}_i(0 \dots k_e), \boldsymbol{W}_i(0 \dots k_e))$.

Initialize
| $Poss(z_i(0), f_i) = 1 \ \forall (z_i(0), f_i) \in \mathcal{Z}_{0i} \times \mathcal{N}_{f_i}$.
End
For k_e++
| Wait for $(\boldsymbol{v}_i(k_e)/\boldsymbol{w}_i(k_e))$.
| **For** *all combinations of* f_i
| | **For** *all combinations of* $z_i(k_e+1)$
| | | Do prediction by applying eqns. (6.2) and (6.3).
| | **End**
| | Do projection by applying eqn. (6.4).
| **End**
| Apply eqn. (6.5) to obtain $\mathcal{F}_i^{dc}(k_e)$.
End

Result: Locally obtained set of possible faults $\mathcal{F}_i^{dc}(k_e)$ for $k_e = 0, 1, \dots$.

The local diagnostic algorithm returns always, i.e. independently of autonomy, the best possible result based on the restricted amount of information.

Theorem 8.
The locally obtained diagnostic result $\mathcal{F}_i^{dc}(k_e)$ obtained by Algorithm 7 is complete and sound

$$\mathcal{F}_i^{dc}(k_e) = \mathcal{F}_i^{\star}(k_e) \tag{6.6}$$

for $k_e = 0, 1, \dots$. □

Proof. The proof of Theorem 8 can be held in analogy to the proof of Theorem 5 and is therefore omitted. ■

Global diagnosis. The problem of gaining a decentrally obtained diagnostic result $\mathcal{F}^{dc}(k_e)$ in the merger from the local diagnostic results can be stated as follows:

Decentralized fault diagnostic problem for nondeterministic asynchronous networks of I/O-automata

Given:	Locally obtained sets of possible faults $\mathcal{F}_i^{dc}(k_e)$, $\forall i \in \{1, \dots, \nu\}$
Find:	Decentrally obtained set of possible faults $\mathcal{F}^{dc}(k_e)$

The straightforward solution to the decentralized fault diagnostic problem is given by building the Cartesian product of the locally obtained sets of possible faults to obtain the decentrally obtained set of possible faults $\mathcal{F}^{dc}(k)$:

$$\mathcal{F}^{dc}(k) = \mathcal{F}_1^{dc}(k) \times \ldots \times \ldots \mathcal{F}_\nu^{dc}(k). \tag{6.7}$$

It contains all tuples of faults that can have occurred up to time k_e. The resulting diagnostic algorithm is given as follows:

Algorithm 8: Decentralized diagnosis of a $\mathcal{NAN}$

Given: Locally obtained sets of possible faults $\mathcal{F}_i^{dc}(k)$ $\forall i \in \{1, \ldots, \nu\}$ for time k_e.
For $k_e + +$
  Wait for $\mathcal{F}_i^{dc}(k_e)$ $\forall i \in \{1, \ldots, \nu\}$.
  Apply eqn. (6.7) to obtain $\mathcal{F}^{dc}(k_e)$.
End
Result: Decentrally obtained set of possible faults $\mathcal{F}^{dc}(k_e)$ for $k_e = 0, 1, \ldots$.

Remark. *The locally obtained sets of possible faults $\mathcal{F}_i^{dc}(k_e)$ are updated in the local diagnostic unit D_i^{dc} and recorded to the merger according to the local generation of new control input and/or output. Reliability of Algorithm 8 can be ensured by inserting the following check for the correct update of* all *locally obtained sets of possible faults within the current time step before calculating the decentrally obtained set of possible faults:*

If $\mathcal{F}_i^{dc}(k_e)$ has not been updated for some $i \in \{1, \ldots, \nu\}$ within the time step use $\mathcal{F}_i^{dc}(k_e - 1)$ to proceed.

Thereby, a blocking of the algorithm due to a possible breakdown of a local diagnostic unit or the communication channel can be prevented. Using the locally obtained set of possible faults obtained in the time step before is suitable as the inclusion $\mathcal{F}_i^{dc}(k-1) \supseteq \mathcal{F}_i^{dc}(k)$ holds for all k [113]. □

The properties of the decentrally obtained set of possible faults $\mathcal{F}^{dc}(k_e)$ are investigated in Section 6.3 in detail.

Preprocessing of the component models. All values of the interconnection signals have to be considered in the prediction step of Algorithm 7 to ensure completeness of the diagnostic result by evaluating the term

$$\bigvee_{\boldsymbol{r}_i(k) \in \mathcal{N}_{\boldsymbol{r}_i}} \bigvee_{\boldsymbol{s}_i(k) \in \mathcal{N}_{\boldsymbol{s}_i}} \chi_{\boldsymbol{L}_i^n}(z_i(k+1), \boldsymbol{w}_i(k), \boldsymbol{r}_i(k), z_i(k), \boldsymbol{v}_i(k), \boldsymbol{s}_i(k), f_i)$$

at each time step k. It has been pointed out in [84] that an off-line elimination of the influence of the interconnection signals on the behavioral relation $\boldsymbol{L}_i^n$ is suitable to reduce the online computation time of the algorithm. This elimination is done by applying

$$\check{\chi}_{\boldsymbol{L}_i^n}(z_i', \boldsymbol{w}_i, z_i, \boldsymbol{v}_i, f_i) = \bigvee_{\boldsymbol{r}_i \in \mathcal{N}_{\boldsymbol{r}_i}} \bigvee_{\boldsymbol{s}_i \in \mathcal{N}_{\boldsymbol{s}_i}} \chi_{\boldsymbol{L}_i^n}(z_i', \boldsymbol{w}_i, \boldsymbol{r}_i, z_i, \boldsymbol{v}_i, \boldsymbol{s}_i, f_i) \tag{6.8}$$

to the characteristic function $\chi_{\boldsymbol{L}_i^n}$ of the behavioral relation of each component model. The obtained characteristic function $\check{\chi}_{\boldsymbol{L}_i^n}$[1] is used to improve the result of the prediction step

$$Poss(z_i(k+1), f_i, \boldsymbol{V}_i(0\dots k), \boldsymbol{W}_i(0\dots k)) \tag{6.9}$$
$$= \bigvee_{z_i(k): Poss(z_i(k), f_i, \boldsymbol{V}_i(0\dots k-1), \boldsymbol{W}_i(0\dots k-1))=1} \check{\chi}_{\boldsymbol{L}_i^n}(z_i(k+1), \boldsymbol{w}_i(k), z_i(k), \boldsymbol{v}_i(k), f_i)$$

and its initialization

$$Poss(z_i(1), f_i, \boldsymbol{v}_i(0), \boldsymbol{w}_i(0)) \tag{6.10}$$
$$= \bigvee_{z_i(0): Poss(z_i(0), f_i)=1} \check{\chi}_{\boldsymbol{L}_i^n}(z_i(1), \boldsymbol{w}_i(0), z_i(0), \boldsymbol{v}_i(0), f_i).$$

Replacing eqns. (6.2) and (6.3) in Algorithm 7 by the new prediction step reduces the computation time but does not alter the diagnostic result because the input/output behavior of the components remains unchanged.

6.3 Properties of the decentralized diagnostic method

The properties of the decentrally obtained diagnostic result obtained by Algorithms 7 and 8 are studied in this section. The general case allows for interaction between the components and comes along with a degradation of the decentrally obtained diagnostic result compared to the ideal one (Section 6.3.1). A lack of soundness is investigated in Section 6.3.2. The influence of state-dependent autonomy is analyzed in Section 6.3.3. It ensures equality of the decentrally obtained diagnostic result with the ideal one. Hence, the best diagnostic result can be obtained with the most restrictive set-up for diagnosis.

6.3.1 General case

Interaction between all components results in general in the mutual dependency of the behavior of the components on each other. This dependency is neglected in decentralized diagnosis by considering only the control signals. Thus, the components are considered as independent. Therefore, the decentrally obtained diagnostic result degrades at all:

Theorem 9.
The diagnostic result $\mathcal{F}^{dc}(k_e)$ obtained by Algorithms 7 and 8 is complete but not sound

$$\mathcal{F}^{dc}(k_e) \supseteq \mathcal{F}^{c}(k_e) = \mathcal{F}^{\star}(k_e) \tag{6.11}$$

for $k_e = 0, 1, \dots$. □

[1] The characteristic function resulting from the elimination of the coupling signals is symbolized by $\check{\chi}_{\boldsymbol{L}_i^n}$. It is not to be confused with the characteristic function $\bar{\chi}_{\boldsymbol{L}_i^n}$ in the case of autonomy and the characteristic function $\chi_{\tilde{\boldsymbol{L}}_i^n}$ of an I/O-automaton without coupling signals. In general, $\check{\chi}_{\boldsymbol{L}_i^n} \neq \bar{\chi}_{\boldsymbol{L}_i^n}$ holds.

Proof. Equation (6.11) can be inserted into the definition of completeness (2.9) to obtain

$$\boldsymbol{f} \in \mathcal{F}^{dc}(k_e) \Leftarrow \boldsymbol{B}(k_e) \in \mathcal{B}_{\tilde{\mathcal{N}}}(\boldsymbol{f}) = \mathcal{B}_{\mathcal{NAN}}(\boldsymbol{f}). \tag{6.12}$$

It has to be shown that this relation holds for each $0 \leq k \leq k_e$ to prove the theorem.

The right-hand side of eqn. (6.12) can be rewritten in analogy to the right-hand side of eqn. (5.12) in terms of the equivalent I/O-automaton $\tilde{\mathcal{N}}$ of the network as

$$\bigvee_{\boldsymbol{z}(k_e+1)} \bigvee_{\boldsymbol{z}(k_e)} \cdots \bigvee_{\boldsymbol{z}(0):Poss(\boldsymbol{z}(0),\boldsymbol{f})=1} \bigwedge_{k=0}^{k_e} \chi_{\tilde{\boldsymbol{L}}^n}(\boldsymbol{z}(k+1), \boldsymbol{w}(k), \boldsymbol{z}(k), \boldsymbol{v}(k), \boldsymbol{f}) = 1 \tag{6.13}$$

and with the composition rule (3.79) in terms of the interacting component models as

$$\bigvee_{\boldsymbol{z}(k_e+1)} \bigvee_{\boldsymbol{z}(k_e)} \cdots \bigvee_{\boldsymbol{z}(0):Poss(\boldsymbol{z}(0),\boldsymbol{f})=1} \bigwedge_{k=0}^{k_e} \tag{6.14}$$

$$\left(\bigvee_{\boldsymbol{r}(k)} \bigwedge_{i=1}^{\nu} \chi_{\boldsymbol{L}_i^n}(z_i(k+1), \boldsymbol{w}_i(k), \boldsymbol{r}_i(k), z_i(k), \boldsymbol{v}_i(k), \boldsymbol{k}_i^T \cdot \boldsymbol{r}(k), f_i) \right) = 1.$$

The condition on the left-hand side of eqn. (6.12) can be rewritten using eqn. (6.7) to obtain the relation

$$\boldsymbol{f} = (f_1, \ldots, f_\nu)^T \in \mathcal{F}^{dc}(k_e) \Leftrightarrow \bigwedge_{i=1}^{\nu} f_i \in \mathcal{F}_i^{dc}(k_e) \tag{6.15}$$

and further transformed into

$$\boldsymbol{f} = (f_1, \ldots, f_\nu)^T \in \mathcal{F}^{dc}(k_e) \Leftrightarrow \bigwedge_{i=1}^{\nu} Poss(f_i, \boldsymbol{V}_i(0 \ldots k_e), \boldsymbol{W}_i(0 \ldots k_e)) = 1 \tag{6.16}$$

using eqn. (6.4).

Thus, the inequality

$$\bigwedge_{i=1}^{\nu} Poss(f_i, \boldsymbol{V}_i(0 \ldots k_e), \boldsymbol{W}_i(0 \ldots k_e)) \tag{6.17}$$

$$\geq \bigvee_{\boldsymbol{z}(k_e+1)} \bigvee_{\boldsymbol{z}(k_e)} \cdots \bigvee_{\boldsymbol{z}(0):Poss(\boldsymbol{z}(0),\boldsymbol{f})=1} \bigwedge_{k=0}^{k_e}$$

$$\left(\bigvee_{\boldsymbol{r}(k)} \bigwedge_{i=1}^{\nu} \chi_{\boldsymbol{L}_i^n}(z_i(k+1), \boldsymbol{w}_i(k), \boldsymbol{r}_i(k), z_i(k), \boldsymbol{v}_i(k), \boldsymbol{k}_i^T \cdot \boldsymbol{r}(k), f_i) \right) = 1$$

has to be proven for each $0 \leq k \leq k_e$ to prove the theorem. This proof is held by induction.

Induction basis. For $k_e = 0$, inserting eqn. (6.3) into eqn. (6.4) yields in conjunction with eqn. (6.16)

$$\bigwedge_{i=1}^{\nu} Poss(f_i, \boldsymbol{v}_i(0), \boldsymbol{w}_i(0)) = \bigwedge_{i=1}^{\nu} \bigvee_{z_i(1)} Poss(z_i(1), f_i, \boldsymbol{v}_i(0), \boldsymbol{w}_i(0)) \tag{6.18}$$

$$= \bigwedge_{i=1}^{\nu} \bigvee_{z_i(1)} \bigvee_{z_i(0):Poss(z_i(0),f_i)=1} \bigvee_{\boldsymbol{r}_i(0)} \bigvee_{\boldsymbol{s}_i(0)} \chi_{\boldsymbol{L}_i^n}(z_i(1), \boldsymbol{w}_i(0), \boldsymbol{r}_i(0), z_i(0), \boldsymbol{v}_i(0), \boldsymbol{s}_i(0), f_i) \tag{6.19}$$

$$= \bigvee_{\boldsymbol{z}(1)} \bigvee_{\boldsymbol{z}(0): \bigwedge_{i=1}^{\nu} Poss(z_i(0),f_i)} \bigvee_{\boldsymbol{r}(0)} \bigvee_{\boldsymbol{s}(0)} \bigwedge_{i=1}^{\nu} \chi_{\boldsymbol{L}_i^n}(z_i(1), \boldsymbol{w}_i(0), \boldsymbol{r}_i(0), z_i(0), \boldsymbol{v}_i(0), \boldsymbol{s}_i(0), f_i) \tag{6.20}$$

$$\geq \bigvee_{\boldsymbol{z}(1)} \bigvee_{\boldsymbol{z}(0):Poss(\boldsymbol{z}(0),\boldsymbol{f})=1} \bigvee_{\boldsymbol{r}(0)} \bigwedge_{i=1}^{\nu} \chi_{\boldsymbol{L}_i^n}(z_i(1), \boldsymbol{w}_i(0), \boldsymbol{r}_i(0), z_i(0), \boldsymbol{v}_i(0), \boldsymbol{k}_i^T \cdot \boldsymbol{r}(0), f_i) \tag{6.21}$$

where the relation $\bigwedge_{i=1}^{\nu} \bigvee_{z_i(1)} \bigvee_{z_i(0)} \bigvee_{\boldsymbol{r}_i(0)} \bigvee_{\boldsymbol{s}_i(0)} \widehat{=} \bigvee_{\boldsymbol{z}(1)} \bigvee_{\boldsymbol{z}(0)} \bigvee_{\boldsymbol{r}(0)} \bigvee_{\boldsymbol{s}(0)} \bigwedge_{i=1}^{\nu}$ has been used to obtain eqn. (6.20). The "greater than or equal" sign in eqn. (6.21) proves the induction basis because the result is equal to the right-hand side of eqn. (6.17) for $k_e = 0$.

Inductive step. The inductive hypothesis is proposed by assuming that equation (6.17) holds for some $k_e = k_h$

$$\bigwedge_{i=1}^{\nu} Poss(f_i, \boldsymbol{V}_i(0 \dots k_h), \boldsymbol{W}_i(0 \dots k_h)) \tag{6.22}$$

$$\geq \bigvee_{\boldsymbol{z}(k_h+1)} \bigvee_{\boldsymbol{z}(k_h)} \cdots \bigvee_{\boldsymbol{z}(0):Poss(\boldsymbol{z}(0),\boldsymbol{f})=1} \bigwedge_{k=0}^{k_h}$$

$$\left(\bigvee_{\boldsymbol{r}(k)} \bigwedge_{i=1}^{\nu} \chi_{\boldsymbol{L}_i^n}(z_i(k+1), \boldsymbol{w}_i(k), \boldsymbol{r}_i(k), z_i(k), \boldsymbol{v}_i(k), \boldsymbol{k}_i^T \cdot \boldsymbol{r}(k), f_i) \right).$$

It has to be proven that eqn. (6.17) is satisfied for $k_e = k_h + 1$ based on this hypothesis

$$\bigwedge_{i=1}^{\nu} Poss(f_i, \boldsymbol{V}_i(0 \dots k_h + 1), \boldsymbol{W}_i(0 \dots k_h + 1)) \tag{6.23}$$

$$\geq \bigvee_{\boldsymbol{z}(k_h+2)} \bigvee_{\boldsymbol{z}(k_h+1)} \bigvee_{\boldsymbol{z}(k_h)} \cdots \bigvee_{\boldsymbol{z}(0):Poss(\boldsymbol{z}(0),\boldsymbol{f})=1} \bigwedge_{k=0}^{k_h+1}$$

$$\left(\bigvee_{\boldsymbol{r}(k)} \bigwedge_{i=1}^{\nu} \chi_{\boldsymbol{L}_i^n}(z_i(k+1), \boldsymbol{w}_i(k), \boldsymbol{r}_i(k), z_i(k), \boldsymbol{v}_i(k), \boldsymbol{k}_i^T \cdot \boldsymbol{r}(k), f_i) \right).$$

The right-hand side of this equation can be reformulated as

$$\bigvee_{\boldsymbol{z}(k_h+2)} \bigvee_{\boldsymbol{z}(k_h+1)} \bigvee_{\boldsymbol{z}(k_h)} \cdots \bigvee_{\boldsymbol{z}(0):Poss(\boldsymbol{z}(0),\boldsymbol{f})=1} \bigwedge_{k=0}^{k_h+1}$$

$$\left(\bigvee_{\boldsymbol{r}(k)} \bigwedge_{i=1}^{\nu} \chi_{\boldsymbol{L}_i^n}(z_i(k+1), \boldsymbol{w}_i(k), \boldsymbol{r}_i(k), z_i(k), \boldsymbol{v}_i(k), \boldsymbol{k}_i^T \cdot \boldsymbol{r}(k), f_i) \right)$$

$$= \bigvee_{\boldsymbol{z}(k_h+2)} \left(\bigvee_{\boldsymbol{r}(k_h+1)} \bigwedge_{i=1}^{\nu} \right.$$

$$\chi_{\boldsymbol{L}_i^n}(z_i(k_h+2), \boldsymbol{w}_i(k_h+1), \boldsymbol{r}_i(k_h+1), z_i(k_h+1), \boldsymbol{v}_i(k_h+1), \boldsymbol{k}_i^T \cdot \boldsymbol{r}(k_h+1), f_i)\Bigg)$$

$$\wedge \left(\bigvee_{\boldsymbol{z}(k_h+1)} \bigvee_{\boldsymbol{z}(k_h)} \cdots \bigvee_{\boldsymbol{z}(0):Poss(\boldsymbol{z}(0),\boldsymbol{f})=1} \bigwedge_{k=0}^{k_h} \right. \tag{6.24}$$

$$\left. \left(\bigvee_{\boldsymbol{r}(k)} \bigwedge_{i=1}^{\nu} \chi_{\boldsymbol{L}_i^n}(z_i(k+1), \boldsymbol{w}_i(k), \boldsymbol{r}_i(k), z_i(k), \boldsymbol{v}_i(k), \boldsymbol{k}_i^T \cdot \boldsymbol{r}(k), f_i) \right) \right)$$

$$\leq \bigwedge_{i=1}^{\nu} \bigvee_{z_i(k_h+2)} \bigvee_{\boldsymbol{r}_i(k_h+1)} \bigvee_{\boldsymbol{s}_i(k_h+1)}$$

$$\chi_{\boldsymbol{L}_i^n}(z_i(k_h+2), \boldsymbol{w}_i(k_h+1), \boldsymbol{r}_i(k_h+1), z_i(k_h+1), \boldsymbol{v}_i(k_h+1), \boldsymbol{s}_i(k_h+1), f_i) \tag{6.25}$$

$$\wedge \left(\bigwedge_{i=1}^{\nu} Poss(f_i, \boldsymbol{V}_i(0 \ldots k_h), \boldsymbol{W}_i(0 \ldots k_h)) \right)$$

$$= \bigwedge_{i=1}^{\nu} Poss(f_i, \boldsymbol{V}_i(0 \ldots k_h+1), \boldsymbol{W}_i(0 \ldots k_h+1)) \tag{6.26}$$

which proves the inductive step and the theorem. ■

6.3.2 Lack of soundness

The spurious solutions $\boldsymbol{f} \in \mathcal{F}^{spur}(k_e)$ which have been defined in eqn. (2.12) as possible faults that are no fault candidates rely on two facts. First, the inequality

$$\bigvee_{\boldsymbol{r}(k) \in \mathcal{N}_{\boldsymbol{r}}} \bigwedge_{i=1}^{\nu} \chi_{\boldsymbol{L}_i^n}(z_i(k+1), \boldsymbol{w}_i(k), \boldsymbol{r}_i(k), z_i(k), \boldsymbol{v}_i(k), \boldsymbol{k}_i^T \cdot \boldsymbol{r}(k), f) \tag{6.27}$$

$$\leq \bigwedge_{i=1}^{\nu} \bigvee_{\boldsymbol{r}_i(k) \in \mathcal{N}_{\boldsymbol{r}_i}} \bigvee_{\boldsymbol{s}_i(k) \in \mathcal{N}_{\boldsymbol{s}_i}} \chi_{\boldsymbol{L}_i^n}(z_i(k+1), \boldsymbol{w}_i(k), \boldsymbol{r}_i(k), z_i(k), \boldsymbol{v}_i(k), \boldsymbol{s}_i(k), f_i)$$

which has been used to obtain eqn. (6.25) in the proof of Theorem 9 will be analyzed. Its left-hand side is associated with the network of I/O-automata. It yields only the value "1" for $\boldsymbol{r}(k)$ being a fixed-point $\bar{\boldsymbol{r}} \in \mathcal{N}_{\boldsymbol{r}}$ that solves the direct feedback problem. The number of fixed-points is typically smaller than the number of symbols of the interconnection output $\boldsymbol{r}(k)$ given by T.

The right-hand side of eqn. (6.27) corresponds to the independent treatment of the components. The problem of direct feedback does not arise because there are no restrictions on the interconnection signals. Thus, the value "1" is obtained for all fixed-points and may be obtained additionally for the remaining values of the interconnection signals. This circumstance is the first reason for the "less than or equal" sign.

The second reason is associated with the preliminary conditions used in the prediction steps (6.2) and (5.20). The equality of these conditions is guaranteed in the initializing step ($k = 0$)

$$Poss(\boldsymbol{z}(0), \boldsymbol{f}) = \bigwedge_{i=1}^{\nu} Poss(z_i(0), f_i). \tag{6.28}$$

It holds as long as equality is given in eqn. (6.27). If inequality is firstly given at time k_e in eqn. (6.27), inequality holds for all future times $k \geq k_e + 1$ for the preliminary conditions of the next time step $k = k_e + 1$:

$$\begin{aligned} &Poss(\boldsymbol{z}(k), \boldsymbol{f}, \boldsymbol{V}(0 \ldots k-1), \boldsymbol{W}(0 \ldots k-1)) \\ &\leq \bigwedge_{i=1}^{\nu} Poss(z_i(k), f_i, \boldsymbol{V}_i(0 \ldots k-1), \boldsymbol{W}_i(0 \ldots k-1)). \end{aligned} \tag{6.29}$$

These considerations are used for the following statement:

Theorem 10.
The decentrally obtained diagnostic result $\mathcal{F}^{dc}(k)$ obtained by Algorithms 7 and 8 is complete and sound until time $k_e - 1$, if and only if equality is guaranteed in eqn. (6.27) *for* $0 \leq k \leq k_e - 1$*:*

$$\begin{aligned} &\bigvee_{\boldsymbol{r}(k) \in \mathcal{N}_{\boldsymbol{r}}} \bigwedge_{i=1}^{\nu} \chi_{\boldsymbol{L}_i^n}(z_i(k+1), \boldsymbol{w}_i(k), \boldsymbol{r}_i(k), z_i(k), \boldsymbol{v}_i(k), \boldsymbol{k}_i^T \cdot \boldsymbol{r}(k), f) \\ &= \bigwedge_{i=1}^{\nu} \bigvee_{\boldsymbol{r}_i(k) \in \mathcal{N}_{\boldsymbol{r}_i}} \bigvee_{\boldsymbol{s}_i(k) \in \mathcal{N}_{\boldsymbol{s}_i}} \chi_{\boldsymbol{L}_i^n}(z_i(k+1), \boldsymbol{w}_i(k), \boldsymbol{r}_i(k), z_i(k), \boldsymbol{v}_i(k), \boldsymbol{s}_i(k), f_i). \end{aligned} \tag{6.30}$$

A lack of soundness arises from time k_e for all future time $k \geq k_e$, if eqn. (6.27) *holds with "<"- sign.* □

Proof. The proof follows directly from the construction procedure described above. ■

The next section investigates conditions to ensure equality in eqns. (6.27) and (6.29) which result in the disappearance of spurious solutions.

6.3.3 Simplifications resulting from state-dependent autonomy

This section is concerned with the influence of the autonomy conditions on the decentrally obtained diagnostic result. Structural autonomy is not considered at this point because the results are obvious due to the first consequence of Definition 13. The more interesting case arises from state-dependent autonomy defined in Section 4.2.2 because the interconnection signals do no longer have any influence on the behavior of the components. As there is no difference between treating the components as independent or not, equality is ensured in eqns. (6.27) and (6.29). Consequently, spurious solutions cannot occur and the decentrally obtained diagnostic result does not degrade compared to the ideal one:

Theorem 11.
The diagnostic result $\mathcal{F}^{dc}(k_e)$ obtained by Algorithms 7 and 8 is complete and sound

$$\mathcal{F}^{dc}(k_e) = \mathcal{F}^{c}(k_e) = \mathcal{F}^{\star}(k_e) \tag{6.31}$$

for $k_e = 0, 1, \ldots,$ if and only if the network is well-posed and the autonomy condition (4.19)

$$\begin{aligned} &\chi_{\boldsymbol{L}_i^n}(z_i(k_e+1), \boldsymbol{w}_i(k_e), \boldsymbol{r}_i(k_e), z_i(k_e), \boldsymbol{v}_i(k_e), \boldsymbol{s}_i(k_e), f_i) \\ &= \check{\chi}_{\boldsymbol{L}_i^n}(z_i(k_e+1), \boldsymbol{w}_i(k_e), z_i(k_e), \boldsymbol{v}_i(k_e)) \end{aligned} \tag{6.32}$$

is satisfied for all $f_i \in \mathcal{F}_i^{dc}(k_e-1)$ $(i \in \{1, \ldots, \nu\})$ given by eqn. (6.5) and the measurements $(\boldsymbol{v}_i(k_e)/\boldsymbol{w}_i(k_e))$ of the control inputs and outputs of the components. □

Proof. Theorem 11 is proven by showing first that equality is given in eqn. (6.27) under the condition (6.32). Inserting this condition into the left-hand side of eqn. (6.27) yields

$$\begin{aligned} &\bigvee_{\boldsymbol{r}(k)\in\mathcal{N}_{\boldsymbol{r}}} \bigwedge_{i=1}^{\nu} \chi_{\boldsymbol{L}_i^n}(z_i(k+1), \boldsymbol{w}_i(k), \boldsymbol{r}_i(k), z_i(k), \boldsymbol{v}_i(k), \boldsymbol{k}_i^T \cdot \boldsymbol{r}(k), f) \\ &= \bigwedge_{i=1}^{\nu} \check{\chi}_{\boldsymbol{L}_i^n}(z_i(k+1), \boldsymbol{w}_i(k), z_i(k), \boldsymbol{v}_i(k)), \end{aligned} \tag{6.33}$$

only if the network is well-posed. As the characteristic function $\check{\chi}_{\boldsymbol{L}_i^n}$ of the right-hand side of eqn. (6.33) does no longer depend on the interconnection outputs, $\bigvee_{\boldsymbol{r}(k)\in\mathcal{N}_{\boldsymbol{r}}}$ can be rejected. For well-posed networks, the insertion of condition (6.32) into the right-hand side of (6.27) yields

$$\begin{aligned} &\bigwedge_{i=1}^{\nu} \bigvee_{\boldsymbol{r}_i(k)\in\mathcal{N}_{\boldsymbol{r}_i}} \bigvee_{\boldsymbol{s}_i(k)\in\mathcal{N}_{\boldsymbol{s}_i}} \chi_{\boldsymbol{L}_i^n}(z_i(k+1), \boldsymbol{w}_i(k), \boldsymbol{r}_i(k), z_i(k), \boldsymbol{v}_i(k), \boldsymbol{s}_i(k), f_i) \\ &= \bigwedge_{i=1}^{\nu} \check{\chi}_{\boldsymbol{L}_i^n}(z_i(k+1), \boldsymbol{w}_i(k), z_i(k), \boldsymbol{v}_i(k)) \end{aligned} \tag{6.34}$$

where $\bigvee_{\boldsymbol{r}_i(k)\in\mathcal{N}_{\boldsymbol{r}_i}}$ and $\bigvee_{\boldsymbol{s}_i(k)\in\mathcal{N}_{\boldsymbol{s}_i}}$ can be rejected because the characteristic function $\check{\chi}_{\boldsymbol{L}_i^n}$ of the right-hand side of this equation does neither depend on the interconnection output nor on the interconnection input. As the right-hand sides of eqns. (6.33) and (6.34) are equal, the left-hand sides are also identical which proves the equality of eqn. (6.27). The equality of eqn. (6.29) is a direct consequence of eqn. (6.28) and the proof of equality of eqn. (6.27).

This result is applied in the second step in the proof of Theorem 9 to prove that eqn. (6.31) holds. The inequality sign in eqn. (6.17) is replaced by the equality sign. Thus, the improved version of eqn. (6.17) has to be proven. This proof can be held in analogy to the original proof with the difference that equality holds in eqns. (6.21) and (6.25) due to the above held prove of equality in eqns. (6.27) and (6.29). This line of reasoning proves Theorem 11. ■

The design strategy of coupled mechatronical systems may require the interaction of several but not all components to fulfill a desired task depending on the operating point. Thus, assuming state-dependent autonomy at each time step k_e as done in Theorem 11 is not realistic. It is more likely that condition (6.32) holds at *some* time steps k such that equality is sometimes given in eqn. (6.27) and not always. The impact on the decentrally obtained diagnostic result is the temporal stopping of its degradation compared to the centrally obtained one because the same possible faults will be excluded in both settings at that time. The decentrally obtained diagnostic result still remains a superset of the centrally obtained one even in these time steps

$$\mathcal{F}^{dc}(k) \supseteq \mathcal{F}^{c}(k) = \mathcal{F}^{\star}(k) \tag{6.35}$$

due to the inequality of the preliminary conditions given by eqn. (6.29). The temporal stopping of this degradation cannot be formalized within the given setting. The simplifications resulting from state-dependent autonomy will be further investigated in partially coordinated diagnosis introduced in Chapter 7.

6.4 Complexity considerations

The complexity of the centralized and decentralized diagnostic algorithms is investigated in this section in terms of the memory consumption of the model and the computational costs arising from the evaluation of the equations of the algorithms [84].

Memory consumption. The behavioral relation of the equivalent I/O-automaton $\tilde{\mathcal{N}}$ of a asynchronous network of I/O-automata $\mathcal{NAN}$ contains at most

$$\prod_{i=1}^{\nu} N_i^2 \cdot M_i \cdot R_i \cdot S_i \tag{6.36}$$

transitions, where N_i denotes the number of states, $M_i = \prod_{j=1}^{\mu_i} M_i^j$ the number of control inputs, $R_i = \prod_{j=1}^{\rho_i} R_i^j$ the number of control outputs and S_i the number of faults of each component respectively. The maximal number of transitions that need to be stored for the network of I/O-automata is given as

$$\sum_{i=1}^{\nu} N_i^2 \cdot M_i \cdot R_i \cdot P_i \cdot T_i \cdot S_i, \tag{6.37}$$

where $P_i = \prod_{j=1}^{\pi_i} P_i^j$ denotes the number of interconnection inputs and $T_i = \prod_{j=1}^{\tau_i} T_i^j$ the number of interconnection outputs. It can be further reduced by preprocessing the component models using eqn. (6.8) to obtain

$$\sum_{i=1}^{\nu} N_i^2 \cdot M_i \cdot R_i \cdot S_i. \tag{6.38}$$

An upper bound can be given for the memory consumption using the Landau symbol $O(\bullet)$ which denotes the order of magnitude of the consumed memory [46]. The sets of signals are approximated through an upper bound denoted by N, M, R, P, T, S. These considerations result in the following approximation

$$O((N^2 \cdot M \cdot R \cdot S)^\nu) \qquad \text{for the equivalent I/O-automaton,} \tag{6.39}$$

$$O(\nu \cdot N^2 \cdot M \cdot R \cdot P \cdot T \cdot S) \qquad \text{for the original network and} \tag{6.40}$$

$$O(\nu \cdot N^2 \cdot M \cdot R \cdot S) \qquad \text{for the preprocessed network.} \tag{6.41}$$

The memory consumption of the network is significantly lower than that of the equivalent I/O-automaton because it grows linearly with number of components of the network as opposed to the exponential growth of the equivalent I/O-automaton. Thus, the resulting I/O-automaton becomes quickly to large to be handleable even with modern computers. Preprocessing of the component models results in a further reduction of the consumed memory of the network of I/O-automata given as the difference of eqn. (6.40) and eqn. (6.41).

Computational costs. The prediction step in centralized diagnosis requires the calculation of N disjunctions for all combinations of successor states and fault cases. N disjunctions are additionally evaluated in the projection step for each fault case. The amount of arithmetic operations is given for *one time step* as

$$N \cdot N \cdot S + N \cdot S \tag{6.42}$$

which is of order $O(N^2 \cdot S)$.

The local diagnostic algorithm used in the decentralized set-up is of the same order as the centralized diagnostic algorithm. It is used in each local diagnostic unit in parallel. The combination of the locally obtained diagnostic results requires at most the evaluation of

$$\nu \cdot S \tag{6.43}$$

arithmetic operations in the merger where $S = \prod_{i=1}^{\nu} S_i$ holds. It can be neglected in comparison to the cost of the local diagnostic algorithm as it is of linear order $O(\nu \cdot S)$.

Even though the algorithmic costs are equal in the centralized and the decentralized set-up, the quantity of the arguments differs significantly. The number of states of the equivalent I/O-automaton used for centralized diagnosis is given as the product of the number of states of the component models used for decentralized diagnosis, i.e. $N = \prod_{i=1}^{\nu} N_i$ holds. Even if it might be possible to carry out the composition to achieve the equivalent I/O-automaton of a complex system, the centralized diagnostic algorithm would not finish in a reasonable time for large N because of its quadratic order.

6.5 Evaluation of results

The decentralization of the diagnostic task has been considered in this chapter. It comes along with a significant reduction of complexity in terms of the memory needed to store the model and the computational costs given as the number of arithmetic operations used in the diagnostic algorithm. Even though the order of the decentralized diagnostic algorithm is equal to the centralized one, the total amount of operations differs highly due to the quantity of the arguments. Decentralized diagnosis is more *reusable* than the centralized set-up because changes in a component affect only the corresponding model and do not result in re-composing the entire system model. Good *scalability* is given because adding or removing components alters only the merger and, therefore, has no effect on the remaining system. A possible breakdown of one or several local diagnostic units does not result in the total failure of the diagnostic system. Thus, this approach is *reliable*.

The asynchronous motion of components is indicated in this thesis by the explicit use of the empty symbol ε to ensure equality of the length of all sequences of measured control inputs and outputs. The empty symbol is used in the decentralized diagnostic algorithm presented in this chapter as a regular symbol. Hence, the locally obtained diagnostic results are synchronized on the occurrence of events and not on a clock signal like in [84]. Even though there is a difference in way of updating the algorithms, the used equations do not differ for asynchronous and synchronous networks of I/O-automata.

The decentrally obtained diagnostic result obtained by Algorithms 7 and 8 is equal to the ideal diagnostic result in the case of *state-dependent autonomy* for the given measurements. Thus, it yields the best diagnostic result based on the most restrictive information structure. The drawback is a lack of soundness for the *general case* allowing for interaction between the components due to disregarding of the interconnection signals. It has been proven that the decentrally obtained diagnostic result is complete such that the system is never diagnosed as faultless, if a fault has occurred. Consequently, the decentrally obtained diagnostic result is a superset of the centrally obtained one: $\mathcal{F}^{\mathrm{dc}}(k_e) \supseteq \mathcal{F}^{\mathrm{c}}(k_e) = \mathcal{F}^{\bigstar}(k_e)$. The degradation of the decentrally obtained diagnostic result can be stopped at time k, if all components are in state-dependent autonomy because the same possible faults are excluded in the decentralized and the centralized setting. Spurious solutions can be avoided in the case of state-dependent autonomy but not always. Hence, soundness cannot be ensured always in the decentralized diagnostic approach considered in this chapter. The next chapter presents a diagnostic method that takes the interconnection signals into account to guarantee soundness.

Chapter 7

Partially coordinated diagnosis of nondeterministic asynchronous networks of I/O-automata

This chapter develops a diagnostic method that yields the ideal diagnostic result, which has been obtained in the centralized approach, and assures the complexity advantages of the decentralized set-up. The new information structure is motivated in Section 7.1. The resulting partial coordination scheme is presented in Section 7.2. The complexity of the diagnostic algorithm is analyzed in Section 7.3. An example comparing the diagnostic methods presented in this thesis is given in Section 7.4. This chapter closes in Section 7.5 with an evaluation of these approaches.

7.1 Motivation for a new information structure

The decentrally obtained diagnostic result has been proven to be equal to the centrally obtained and the ideal diagnostic result in the case of state-dependent autonomy for the given measurements (Theorem 11) for $0 \leq k \leq k_e$ because the behavior of the components does not depend on each other:

$$\mathcal{F}^{dc}(k) = \mathcal{F}^{c}(k) = \mathcal{F}^{\star}(k). \tag{7.1}$$

The interconnection signals may be neglected without any impact on the diagnostic result. A lack of soundness arises firstly, as specified in Theorem 10, if the interrelations with other components are to be regarded. The corresponding degradation of the decentrally obtained diagnostic result to a superset of the centrally obtained and the ideal one results in the existence of spurious solutions given by eqn. (2.12). The relation

$$\mathcal{F}^{dc}(k) \supseteq \mathcal{F}^{c}(k) = \mathcal{F}^{\star}(k) \tag{7.2}$$

known from Theorem 9 is inherent to the information structure and neither to the diagnostic algorithm nor to the used type of model.

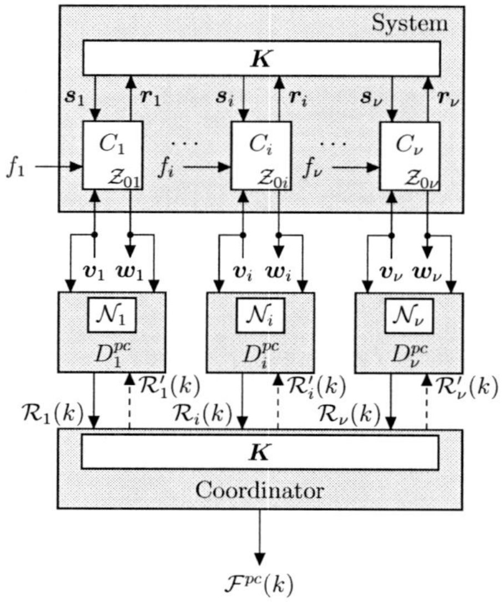

Figure 7.1: Partially coordinated diagnosis of an interconnected discrete-event system

The only way to obtain the ideal diagnostic result while assuring the complexity advantages of the decentralization is to alter the information structure by estimating the possible values of the interconnection signals instead of considering all of them. These locally estimated values are improved in a new component called *coordinator* which calculates the coordinated diagnostic result $\mathcal{F}^{pc}(k)$. It has been originally introduced in [84] for synchronous networks of I/O-automata based on relational algebra. The bidirectional information structure shown in Fig. 7.1 has been proven therein to yield the ideal diagnostic result $\mathcal{F}^\star(k)$ at all times k. Its application results in the following changes of the decentralized set-up:

1. The local diagnostic unit D_i^{pc} of a component C_i observes the interconnection inputs $\boldsymbol{s}_i(k)$ and outputs $\boldsymbol{r}_i(k)$ in addition to the fault f_i. The locally obtained sets of possible pairs $\mathcal{R}_i(k) = \{(\boldsymbol{s}_i(k), \boldsymbol{r}_i(k), f_i)\}$ are sent to the coordinator.

2. The coordinator compares these observations based on the interconnection model $\boldsymbol{K}$ and removes contradictions from the sets $\mathcal{R}_i(k)$. It calculates the coordinated diagnostic result $\mathcal{F}^{pc}(k)$ of the system.

3. The improved locally obtained sets of possible pairs $\mathcal{R}_i'(k) = \{(\boldsymbol{s}_i(k), \boldsymbol{r}_i(k), f_i)\}$ are sent to the local diagnostic units to update the locally obtained results.

This concept will be extended in the following to cope with asynchronous networks of I/O-automata introduced in Chapter 3 in two ways. The formalism used in [84] is adapted to handle models that are based on behavioral relations instead of relational algebra. The more interesting issue deals with the question on whether to coordinate or not, if the components are in state-dependent autonomy to obtain the ideal diagnostic result $\mathcal{F}^{\star}(k)$ and, therefore, to avoid spurious solutions with the less possible effort.

7.2 Solution of the partially coordinated diagnostic problem

Partially coordinated diagnosis consists of ν local diagnostic unit D_i^{pc} and a central coordinator. Each local diagnostic unit estimates for $0 \leq k \leq k_e$ the locally obtained sets of possible pairs $\mathcal{R}_i(k)$ based on a consistency-test between the component model given in terms of a nondeterministic I/O-automaton $\mathcal{N}_i$ and the local measurements $\tilde{\boldsymbol{B}}_i(k_e) = (\boldsymbol{V}_i(0 \ldots k_e), \boldsymbol{W}_i(0 \ldots k_e))$. The aim of the coordinator is to *remove contradictions* from these sets to calculate the partially coordinated set of possible faults $\mathcal{F}^{pc}(k)$. The improved locally obtained sets of possible pairs $\mathcal{R}_i'(k)$ are sent back to the local diagnostic units to enhance their results. Thus, their is a bidirectional exchange of information between the local diagnostic units and the coordinator. The aim of this approach is to restrict this information flow to a minimum using the simplifications resulting from state-dependent autonomy. This diagnostic problem can be stated in the following way:

	Diagnostic problem for nondeterministic asynchronous networks of I/O-automata
Given:	Initialized nondeterministic asynchronous network of I/O-automata $\mathcal{NAN}$
	Local measurements $\tilde{\boldsymbol{B}}_i(k_e) = (\boldsymbol{V}_i(0 \ldots k_e), \boldsymbol{W}_i(0 \ldots k_e))$
Find:	Partially coordinated set of possible faults $\mathcal{F}^{pc}(k_e)$

The solution to this problem is solved in the *local diagnostic units* and the *centralized coordinator*. The locally obtained results are communicated to the coordinator to improve the result. The result is used to calculate the *partially coordinated set of possible faults* and send back to enhance the locally obtained results. The amount of exchanged data is reduced in the case of state-dependent autonomy.

Local diagnostic units. The local diagnostic Algorithm 7 known from decentralized diagnosis is adapted with the following changes. It starts with the extension of the projection step given by eqn. (6.4):

$$
\begin{aligned}
&Poss(\boldsymbol{s}_i(k), \boldsymbol{r}_i(k), f_i, \boldsymbol{V}_i(0 \ldots k), \boldsymbol{W}_i(0 \ldots k)) \qquad (7.3)\\
&= \bigvee_{z_i(k+1) \in \mathcal{N}_{z_i}} \quad \bigvee_{z_i(k): Poss(z_i(k), f_i, \boldsymbol{V}_i(0 \ldots k-1), \boldsymbol{W}_i(0 \ldots k-1))=1}
\end{aligned}
$$

$$\chi_{\boldsymbol{L}_i^n}(z_i(k+1), \boldsymbol{w}_i(k), \boldsymbol{r}_i(k), z_i(k), \boldsymbol{v}_i(k), \boldsymbol{s}_i(k), f_i).$$

This new projection step determines the possibility of the tuple $(\boldsymbol{s}_i(k), \boldsymbol{r}_i(k), f_i)$ of interconnection inputs, outputs and faults based on the result of the previous prediction step given by all states $z_i(k)$ for which $Poss(z_i(k), f_i, \boldsymbol{V}_i(0\dots k-1), \boldsymbol{W}_i(0\dots k-1)) = 1$ holds under the condition that the sequences of input values $\boldsymbol{V}_i(0\dots k)$ and output values $\boldsymbol{W}_i(0\dots k)$ have been measured. This evaluation has to be carried out independently of state-dependent autonomy given by eqn. (4.19) because the autonomy condition

$$\begin{aligned}
&\chi_{\boldsymbol{L}_i^n}(z_i(k+1), \boldsymbol{w}_i(k), \boldsymbol{r}_i(k), z_i(k), \boldsymbol{v}_i(k), \boldsymbol{s}_i(k), f_i) \\
&\quad \wedge Poss(z_i(k), f_i, \boldsymbol{V}_i(0\dots k-1), \boldsymbol{W}_i(0\dots k-1)) \\
&\quad = \bar{\chi}_{\boldsymbol{L}_i^n}(z_i(k+1), \boldsymbol{w}_i(k), z_i(k), \boldsymbol{v}_i(k)) \wedge Poss(z_i(k), f_i, \boldsymbol{V}_i(0\dots k-1), \boldsymbol{W}_i(0\dots k-1))
\end{aligned} \tag{7.4}$$

guarantees only the independence of the behavior of this component from the *activated* interconnection inputs $\boldsymbol{s}_i \in \mathcal{N}_{\boldsymbol{s}_i}^{act}(z_i)$, outputs $\boldsymbol{r}_i \in \mathcal{N}_{\boldsymbol{r}_i}^{act}(z_i)$ and fault f_i for the actual measurement $(\boldsymbol{v}_i(k), \boldsymbol{w}_i(k))$ and the pairs $(z_i(k), f_i)$ of possible states and faults of the previous step. The coordinator needs to know these values of the interconnection signals to restrict that behavior of the connected components that is defined for some $\boldsymbol{s}_i \in \mathcal{N}_{\boldsymbol{s}_i} \backslash \mathcal{N}_{\boldsymbol{s}_i}^{act}(z_i)$ or $\boldsymbol{r}_i \in \mathcal{N}_{\boldsymbol{r}_i} \backslash \mathcal{N}_{\boldsymbol{r}_i}^{act}(z_i)$ respectively. The locally obtained set of possible pairs, given by its characteristic function

$$\chi_{\mathcal{R}_i(k)}(\boldsymbol{s}_i(k), \boldsymbol{r}_i(k), f_i) = Poss(\boldsymbol{s}_i(k), \boldsymbol{r}_i(k), f_i, \boldsymbol{V}_i(0\dots k), \boldsymbol{W}_i(0\dots k)), \tag{7.5}$$

is sent to the coordinator. The condition for state-dependent autonomy cannot be projected onto the locally obtained set of possible pairs because the dependence on the successor state given in eqn. (7.4) would be rejected. Hence, the coordinator must be "informed" about autonomy by the local diagnostic unit. An extra variable can be used to deliver that piece of information to the coordinator.

No improvements of the result of eqn. (7.5) are possible for well-posed networks, if the autonomy condition (7.4) holds. The locally obtained results are sufficient to evaluate the new prediction step

$$\begin{aligned}
&Poss(z_i(k+1), f_i, \boldsymbol{V}_i(0\dots k), \boldsymbol{W}_i(0\dots k)) \\
&= \bigvee_{z_i(k): Poss(z_i(k), f_i, \boldsymbol{V}_i(0\dots k-1), \boldsymbol{W}_i(0\dots k-1))=1} \\
&\quad \bigvee_{\boldsymbol{r}_i(k)\in\mathcal{N}_{\boldsymbol{r}_i}} \bigvee_{\boldsymbol{s}_i(k)\in\mathcal{N}_{\boldsymbol{s}_i}} \chi_{\boldsymbol{L}_i^n}(z_i(k+1), \boldsymbol{w}_i(k), \boldsymbol{r}_i(k), z_i(k), \boldsymbol{v}_i(k), \boldsymbol{s}_i(k), f_i) \\
&\quad \wedge \chi_{\mathcal{R}_i(k)}(\boldsymbol{s}_i(k), \boldsymbol{r}_i(k), f_i)
\end{aligned} \tag{7.6}$$

where the initial condition $Poss(z_i(0), f_i) = 1 \; \forall (z_i(0), f_i) \in \mathcal{Z}_{0i} \times \mathcal{N}_{f_i}$ is used in the first step $(k = 0)$.

Processing is stopped in the local diagnostic unit D_i^{pc} until the locally obtained set of impossible pairs $\mathcal{R}_i'(k)$ is received from the coordinator, if the autonomy condition (7.4) does not hold. The improved locally obtained set of possible pairs

$$\chi_{\mathcal{R}_i(k)}(\boldsymbol{s}_i(k), \boldsymbol{r}_i(k), f_i) = \chi_{\mathcal{R}_i(k)}(\boldsymbol{s}_i(k), \boldsymbol{r}_i(k), f_i) \wedge \neg\chi_{\mathcal{R}'_i(k)}(\boldsymbol{s}_i(k), \boldsymbol{r}_i(k), f_i) \tag{7.7}$$

is used in the prediction step given by eqn. (7.6). The resulting algorithm is evaluated in each local diagnostic unit D_i^{pc}:

Algorithm 9: Partially coordinated diagnosis of a $\mathcal{NAN}$: Local diagnosis

Given: Initialized nondeterministic I/O-automaton $\mathcal{N}_i$ given by eqn. (3.29),
Local measurements $\tilde{\boldsymbol{B}}_i(k_e) = (\boldsymbol{V}_i(0 \dots k_e), \boldsymbol{W}_i(0 \dots k_e))$.

Initialize
| $Poss(z_i(0), f_i) = 1 \ \forall (z_i(0), f_i) \in \mathcal{Z}_{0i} \times \mathcal{N}_{f_i}$.
End
For $k_e + +$
  Wait for $(\boldsymbol{v}_i(k_e)/\boldsymbol{w}_i(k_e))$.
  For *all combinations of* $\boldsymbol{s}_i(k)$, $\boldsymbol{r}_i(k)$ *and* f_i
  | Do projection by applying eqn. (7.3).
  End
  Calculate locally obtained set of possible pairs by applying eqn. (7.5).
  Send $\mathcal{R}_i(k_e)$ to the coordinator.
  if *state-dependent autonomy is given, i.e. eqn.* (7.4) *holds* **then**
    For *all combinations of* $z_i(k_e + 1)$ *and* f_i
    | Do prediction by applying eqn. (7.6).
    End
  else
    Wait for $\mathcal{R}'_i(k_e)$. If $\mathcal{R}'_i(k_e)$ is not updated, use $\mathcal{R}'_i(k_e) = \emptyset$.
    Apply eqn. (7.7) to improve $\mathcal{R}_i(k_e)$.
    For *all combinations of* $z_i(k_e + 1)$ *and* f_i
    | Do prediction by applying eqn. (7.6).
    End
  end
End

Result: Locally obtained set of possible pairs $\mathcal{R}_i(k_e)$ for $k_e = 0, 1, \dots$.

Remark. *The test*

If $\mathcal{R}'_i(k_e)$ *is not updated, use* $\mathcal{R}'_i(k_e) = \emptyset$

has been inserted in Algorithm 9 to prevent a blocking due to a possible breakdown of the communication with the coordinator. Using $\mathcal{R}'_i(k_e) = \emptyset$ *is the safe side because not refining the locally obtained set of possible pairs results only in spurious solutions and does not exclude fault candidates wrongly.* □

The Algorithm 9 returns in the locally obtained set of possible pairs $\mathcal{R}_i(k_e)$ for each time k_e the locally obtained set of possible faults $\mathcal{F}_i^{dc}(k_e)$ obtained by Algorithm 7 of the decentralized set-up.

Coordinator. The aim of the coordinator is to remove contradictions from the locally obtained sets of possible pairs to gain the ideal diagnostic result. As these contradictions cannot arise in *total autonomy*, the coordinator checks first, if the condition (7.4) holds for *all* local diagnostic results. This piece of information is delivered by the local diagnostic units. The possible faults $\boldsymbol{f}$ can be obtained in this case from $\mathcal{R}_i(k)$ without any improvements by

$$\boxed{Poss(\boldsymbol{f},k) = \bigwedge_{i=1}^{\nu} \bar{\chi}_{\mathcal{R}_i(k)}(f_i) \wedge Poss(\boldsymbol{f},k-1).} \tag{7.8}$$

The possible faults $\boldsymbol{f}$ of the previous step have to be used in the current step because some of the possible combinations of locally obtained faults might have been rejected by coordination from the set $\mathcal{F}^{pc}(k-1)$. The relation

$$Poss(\boldsymbol{f},k) = \bigwedge_{i=1}^{\nu} Poss(f_i,k) \wedge Poss(\boldsymbol{f},k-1) \tag{7.9}$$

shows that these dependencies cannot be stored locally because removing the last term of the right-hand side results in a "less-than-or-equal" sign. Equation (7.8) is initialized with $Poss(\boldsymbol{f},-1) = 1 \; \forall \boldsymbol{f} \in \mathcal{N}_{\boldsymbol{f}}$. The partially coordinated diagnostic result is given as

$$\mathcal{F}^{pc}(k) = \left\{ \boldsymbol{f} = (f_1,\ldots,f_\nu)^T | Poss(\boldsymbol{f},k) = 1 \right\}. \tag{7.10}$$

No data is sent back to the local diagnostic units.

If *at least one component is not in state-dependent autonomy*, the coordination scheme known from [84] is adapted to remove contradictions from the locally obtained set of possible pairs. The resulting coordinated set of pairs $\mathcal{J}(k) = \{(\boldsymbol{s}(k),\boldsymbol{r}(k),\boldsymbol{f})\}$ is given by its characteristic function

$$\chi_{\mathcal{J}(k)}(\boldsymbol{s}(k),\boldsymbol{r}(k),\boldsymbol{f}) = \bigvee_{\boldsymbol{r}(k)\in\mathcal{N}_{\boldsymbol{r}}} \bigwedge_{i=1}^{\nu} \chi_{\mathcal{R}_i(k)}(\boldsymbol{k}_i^T \cdot \boldsymbol{r}_i(k), \boldsymbol{r}_i(k), f_i) \tag{7.11}$$

where the coupling model $\boldsymbol{K}$ given by eqn. (3.61) and the reduced composition rule (3.79) have been used. The locally obtained set of impossible pairs $\mathcal{R}'_i(k)$, given by

$$\chi_{\mathcal{R}'_i(k)}(\boldsymbol{s}_i(k),\boldsymbol{r}_i(k),f_i) = \chi_{\mathcal{R}_i(k)}(\boldsymbol{s}_i(k),\boldsymbol{r}_i(k),f_i) \wedge \neg\chi_{\mathcal{J}(k)}(\boldsymbol{s}(k),\boldsymbol{r}(k),\boldsymbol{f}), \tag{7.12}$$

is sent to the local diagnostic unit D_i^{pc} to improve the locally obtained result. No data is sent to the local diagnostic units whose components are in state-dependent autonomy because their behavior does not depend on the interconnection signals. The possible faults $\boldsymbol{f}$

$$\boxed{Poss(\boldsymbol{f},k) = \left(\bigvee_{\boldsymbol{r}(k)\in\mathcal{N}_{\boldsymbol{r}}} \bigvee_{\boldsymbol{s}(k)\in\mathcal{N}_{\boldsymbol{s}}} \chi_{\mathcal{J}(k)}(\boldsymbol{s}(k),\boldsymbol{r}(k),\boldsymbol{f}) \right) \wedge Poss(\boldsymbol{f},k-1)} \tag{7.13}$$

are obtained by the result of eqn. (7.11) and used in eqn. (7.10) to get the partially coordinated set of possible faults $\mathcal{F}^{pc}(k)$. The resulting algorithm is used in the coordinator:

Algorithm 10: Partially coordinated diagnosis of a $\mathcal{NAN}$: Global coordination

Given: Coupling model $\boldsymbol{K}$ and set of faults $\mathcal{N}_{\boldsymbol{f}}$,
Locally obtained sets of possible pairs $\mathcal{R}_i(k_e)$ $\forall i \in \{1, \dots, \nu\}$ for time k_e.

Initialize
| $Poss(\boldsymbol{f}, -1) = 1 \; \forall \boldsymbol{f} \in \mathcal{N}_{\boldsymbol{f}}$.
End
For $k_e + +$
  Wait for $\mathcal{R}_i(k_e)$ $\forall i \in \{1, \dots, \nu\}$. If $\mathcal{R}_i(k_e)$ is not updated, use
  $\mathcal{R}_i(k_e) = \mathcal{N}_{\boldsymbol{s}_i} \times \mathcal{N}_{\boldsymbol{r}_i} \times \mathcal{N}_{f_i}$.
  if *state-dependent autonomy is given* $\forall i \in \{1, \dots, \nu\}$*, i.e. eqn.* (7.4) *holds* **then**
    | Apply eqns. (7.8) and (7.10) to obtain $\mathcal{F}^{pc}(k_e)$.
  else
    Do coordination by applying equation (7.11) .
    Apply eqns. (7.13) and (7.10) to obtain $\mathcal{F}^{pc}(k_e)$.
    For $i = 1 : \nu$
      if *eqn.* (7.4) *does not hold* **then**
        Apply eqn. (7.12) to calculate $\mathcal{R}_i'(k_e)$.
        Send $\mathcal{R}_i'(k_e)$ to the local diagnostic unit D_i^{pc}.
      end
    End
  end
End

Result: Partially coordinated set of possible faults $\mathcal{F}^{pc}(k_e)$ and locally obtained sets of impossible pairs $\mathcal{R}_i'(k_e)$ for $k_e = 0, 1, \dots$.

Remark. *The test*

If $\mathcal{R}_i(k_e)$ *is not updated, use* $\mathcal{R}_i(k_e) = \mathcal{N}_{\boldsymbol{s}_i} \times \mathcal{N}_{\boldsymbol{r}_i} \times \mathcal{N}_{f_i}$

has been inserted in Algorithm 10 to prevent a blocking due to a possible breakdown of the communication with the according local diagnostic unit. By using $\mathcal{R}_i(k_e) = \mathcal{N}_{\boldsymbol{s}_i} \times \mathcal{N}_{\boldsymbol{r}_i} \times \mathcal{N}_{f_i}$*, no fault candidate is excluded wrongly but faults that are no fault candidates may be taken into account.* □

Remark. *The locally obtained sets of possible pairs must always be coordinated, if all components are in state-dependent autonomy and the network is* not well-posed *because the condition* (7.4) *guarantees only the independence of a component from its neighbors. This condition cannot ensure the existence of* at least one *fixed-point solving eqn.* (4.7)*. The existence of such a interconnection output must be ensured by either assuming that the network is well-posed or by always coordinating the locally obtained diagnostic results.*
□

Partially coordinated diagnosis returns always the best possible diagnostic result:

Theorem 12.
The diagnostic result $\mathcal{F}^{pc}(k_e)$ obtained by Algorithms 9 and 10 is complete and sound

$$\mathcal{F}^{pc}(k_e) = \mathcal{F}^{c}(k_e) = \mathcal{F}^{\star}(k_e) \tag{7.14}$$

for $k_e = 0, 1, \ldots,$ if and only if the network is well-posed. □

Proof. The proof of Theorem 12 is held in analogy to the proof of Theorem 9. Equation (7.14) can be inserted into the definition of consistency (2.7) to obtain

$$\boldsymbol{f} \in \mathcal{F}^{pc}(k_e) \Leftrightarrow \boldsymbol{B}(k_e) \in \mathcal{B}_{\bar{\mathcal{N}}}(\boldsymbol{f}) = \mathcal{B}_{\mathcal{NAN}}(\boldsymbol{f}). \tag{7.15}$$

It has to be shown that this relation holds for each $0 \le k \le k_e$ in the case of state-dependent autonomy and in the general case allowing for interaction between the components to prove the theorem.

State-dependent autonomy. The right-hand side of eqn. (7.15) can be rewritten in analogy to the transformation of the right-hand side of eqn. (6.12) to obtain (6.14) which yields with the condition (7.4) the relation

$$\bigvee_{\boldsymbol{z}(k_e+1)} \bigvee_{\boldsymbol{z}(k_e)} \cdots \bigvee_{\boldsymbol{z}(0): Poss(\boldsymbol{z}(0), \boldsymbol{f})=1} \bigwedge_{k=0}^{k_e} \left(\bigwedge_{i=1}^{\nu} \chi_{\boldsymbol{L}_i^n}(z_i(k+1), \boldsymbol{w}_i(k), z_i(k), \boldsymbol{v}_i(k), f_i) \right) = 1. \tag{7.16}$$

As the behavioral relation in eqn. (7.16) does no longer depend on the interconnection outputs, $\bigvee_{\boldsymbol{r}(k) \in \mathcal{N}_r}$ can be rejected.

The condition on the left-hand side of eqn. (7.15) can be transformed by inserting eqns. (7.5) and (7.4) into eqn. (7.8). The result is used in eqn. (7.10) to obtain

$$\begin{aligned} &\boldsymbol{f} = (f_1, \ldots, f_\nu)^T \in \mathcal{F}^{pc}(k_e) \\ &\Leftrightarrow \bigwedge_{i=1}^{\nu} Poss(\boldsymbol{s}_i(k), \boldsymbol{r}_i(k), f_i, \boldsymbol{V}_i(0 \ldots k), \boldsymbol{W}_i(0 \ldots k)) \wedge Poss(\boldsymbol{f}, k-1) = 1. \end{aligned} \tag{7.17}$$

Thus, the equality

$$\begin{aligned} &\bigwedge_{i=1}^{\nu} Poss(\boldsymbol{s}_i(k), \boldsymbol{r}_i(k), f_i, \boldsymbol{V}_i(0 \ldots k), \boldsymbol{W}_i(0 \ldots k)) \wedge Poss(\boldsymbol{f}, k-1) \\ &= \bigvee_{\boldsymbol{z}(k_e+1)} \bigvee_{\boldsymbol{z}(k_e)} \cdots \bigvee_{\boldsymbol{z}(0): Poss(\boldsymbol{z}(0), \boldsymbol{f})=1} \bigwedge_{k=0}^{k_e} \left(\bigwedge_{i=1}^{\nu} \chi_{\boldsymbol{L}_i^n}(z_i(k+1), \boldsymbol{w}_i(k), z_i(k), \boldsymbol{v}_i(k), f_i) \right) \end{aligned} \tag{7.18}$$

has to be proven for each $0 \le k \le k_e$ to prove the theorem in the case of autonomy. This proof is held by induction.

Induction basis. For $k_e = 0$, the initial condition $Poss(z_i(0), f_i) = 1 \; \forall (z_i(0), f_i) \in \mathcal{Z}_{0i} \times \mathcal{N}_{f_i}$ and the autonomy condition (7.4) are inserted into eqn. (7.3). The first term of the left-hand side of eqn. (7.18) is substituted with that result to obtain

$$\bigwedge_{i=1}^{\nu} Poss(\boldsymbol{s}_i(0), \boldsymbol{r}_i(0), f_i, \boldsymbol{v}_i(0), \boldsymbol{w}_i(0)) \wedge Poss(\boldsymbol{f}, -1) \tag{7.19}$$

$$= \bigwedge_{i=1}^{\nu} \left(\bigvee_{z_i(1)} \bigvee_{z_i(0): Poss(z_i(0), f_i)=1} \bar{\chi}_{\boldsymbol{L}_i^n}(z_i(1), \boldsymbol{w}_i(0), z_i(0), \boldsymbol{v}_i(0), f_i) \right) \wedge Poss(\boldsymbol{f}, -1)$$

$$= \bigvee_{\boldsymbol{z}(1)} \bigvee_{\boldsymbol{z}(0): \bigwedge_{i=1}^{\nu} Poss(z_i(0), f_i)=1} \left(\bigwedge_{i=1}^{\nu} \bar{\chi}_{\boldsymbol{L}_i^n}(z_i(1), \boldsymbol{w}_i(0), z_i(0), \boldsymbol{v}_i(0), f_i) \right) \wedge Poss(\boldsymbol{f}, -1)$$

$$= \bigvee_{\boldsymbol{z}(1)} \bigvee_{\boldsymbol{z}(0): Poss(\boldsymbol{z}(0), \boldsymbol{f})=1} \left(\bigwedge_{i=1}^{\nu} \bar{\chi}_{\boldsymbol{L}_i^n}(z_i(1), \boldsymbol{w}_i(0), z_i(0), \boldsymbol{v}_i(0), f_i) \right)$$

which is the right-hand side of eqn. (7.18) for $k_e = 0$. These transformations prove the induction basis for the case of autonomy.

Inductive step. The inductive hypothesis is proposed by assuming that eqn. (7.18) holds for some $k_e = k_h$ and is satisfied for $k_e = k_h + 1$ based on this hypothesis

$$\bigwedge_{i=1}^{\nu} Poss(\boldsymbol{s}_i(k_h+1), \boldsymbol{r}_i(k_h+1), f_i, \boldsymbol{V}_i(0 \dots k_h+1), \boldsymbol{W}_i(0 \dots k_h+1)) \tag{7.20}$$

$$\wedge \, Poss(\boldsymbol{f}, k_h)$$

$$= \bigvee_{\boldsymbol{z}(k_h+2)} \bigvee_{\boldsymbol{z}(k_h+1)} \bigvee_{\boldsymbol{z}(k_h)} \cdots \bigvee_{\boldsymbol{z}(0): Poss(\boldsymbol{z}(0), \boldsymbol{f})=1}$$

$$\bigwedge_{k=0}^{k_h+1} \left(\bigwedge_{i=1}^{\nu} \bar{\chi}_{\boldsymbol{L}_i^n}(z_i(k+1), \boldsymbol{w}_i(k), z_i(k), \boldsymbol{v}_i(k), f_i) \right).$$

The right-hand side of this equation can be reformulated as

$$\bigvee_{\boldsymbol{z}(k_h+2)} \bigvee_{\boldsymbol{z}(k_h+1)} \bigvee_{\boldsymbol{z}(k_h)} \cdots \bigvee_{\boldsymbol{z}(0): Poss(\boldsymbol{z}(0), \boldsymbol{f})=1} \bigwedge_{k=0}^{k_h+1} \left(\bigwedge_{i=1}^{\nu} \bar{\chi}_{\boldsymbol{L}_i^n}(z_i(k+1), \boldsymbol{w}_i(k), z_i(k), \boldsymbol{v}_i(k), f_i) \right)$$

$$= \bigvee_{\boldsymbol{z}(k_h+2)} \left(\bigwedge_{i=1}^{\nu} \bar{\chi}_{\boldsymbol{L}_i^n}(z_i(k_h+2), \boldsymbol{w}_i(k_h+1), z_i(k_h+1), \boldsymbol{v}_i(k_h+1), f_i) \right) \tag{7.21}$$

$$\wedge \bigvee_{\boldsymbol{z}(k_h+1)} \bigvee_{\boldsymbol{z}(k_h)} \cdots \bigvee_{\boldsymbol{z}(0): Poss(\boldsymbol{z}(0), \boldsymbol{f})=1} \bigwedge_{k=0}^{k_h} \left(\bigwedge_{i=1}^{\nu} \bar{\chi}_{\boldsymbol{L}_i^n}(z_i(k+1), \boldsymbol{w}_i(k), z_i(k), \boldsymbol{v}_i(k), f_i) \right)$$

$$= \bigvee_{\boldsymbol{z}(k_h+2)} \tag{7.22}$$

$$\left(\bigwedge_{i=1}^{\nu} \chi_{\boldsymbol{L}_i^n}(z_i(k_h+2), \boldsymbol{w}_i(k_h+1), \boldsymbol{r}_i(k_h+1), z_i(k_h+1), \boldsymbol{v}_i(k_h+1), \boldsymbol{s}(k_h+1), f_i) \right)$$

$$\wedge \bigvee_{\boldsymbol{z}(k_h+1)} \bigvee_{\boldsymbol{z}(k_h)} \cdots \bigvee_{\boldsymbol{z}(0):\bigwedge_{i=1}^{\nu} Poss(z_i(0),f_i)=1} \bigwedge_{k=0}^{k_h}$$

$$\left(\bigwedge_{i=1}^{\nu} \chi_{\boldsymbol{L}_i^n}(z_i(k+1), \boldsymbol{w}_i(k), \boldsymbol{r}_i(k), z_i(k), \boldsymbol{v}_i(k), \boldsymbol{s}_i(k), f_i)\right) \wedge Poss(\boldsymbol{f}, k_h)$$

$$= \bigwedge_{i=1}^{\nu} \bigvee_{z_i(k_h+2)} \chi_{\boldsymbol{L}_i^n}(z_i(k_h+2), \boldsymbol{w}_i(k_h+1), \boldsymbol{r}_i(k_h+1), z_i(k_h+1), \boldsymbol{v}_i(k_h+1), \boldsymbol{s}(k_h+1), f_i)$$

$$\wedge \left(\bigwedge_{i=1}^{\nu} Poss(\boldsymbol{s}_i(k_h), \boldsymbol{r}_i(k_h), f_i, \boldsymbol{V}_i(0\dots k_h), \boldsymbol{W}_i(0\dots k_h))\right) \wedge Poss(\boldsymbol{f}, k_h) \tag{7.23}$$

$$= \bigwedge_{i=1}^{\nu} Poss(\boldsymbol{s}_i(k_h+1), \boldsymbol{r}_i(k_h+1), f_i, \boldsymbol{V}_i(0\dots k_h+1), \boldsymbol{W}_i(0\dots k_h+1)) \tag{7.24}$$

$$\wedge\, Poss(\boldsymbol{f}, k_h)$$

which proves the inductive step for the case of autonomy.

General case. The right-hand side of eqn. (7.15) can be rewritten in analogy to the transformation of the right-hand side of eqn. (6.12) to obtain (6.14).

The condition on the left-hand side of eqn. (7.15) can be transformed by inserting eqn. (7.5) into eqn. (7.11) and further into eqn. (7.13). The result is used in (7.10) to obtain the relation

$$\boldsymbol{f} = (f_1, \dots, f_\nu)^T \in \mathcal{F}^{pc}(k_e) \tag{7.25}$$

$$\Leftrightarrow \bigvee_{\boldsymbol{r}(k)\in\mathcal{N}_{\boldsymbol{r}}} \bigwedge_{i=1}^{\nu} Poss(\boldsymbol{k}_i^T \cdot \boldsymbol{r}_i(k), \boldsymbol{r}_i(k), f_i, \boldsymbol{V}_i(0\dots k), \boldsymbol{W}_i(0\dots k)) \wedge Poss(\boldsymbol{f}, k-1) = 1.$$

Thus, the equality

$$\bigvee_{\boldsymbol{r}(k_e)\in\mathcal{N}_{\boldsymbol{r}}} \bigwedge_{i=1}^{\nu} Poss(\boldsymbol{k}_i^T \cdot \boldsymbol{r}_i(k_e), \boldsymbol{r}_i(k_e), f_i, \boldsymbol{V}_i(0\dots k_e), \boldsymbol{W}_i(0\dots k_e)) \wedge Poss(\boldsymbol{f}, k-1)$$

$$= \bigvee_{\boldsymbol{z}(k_e+1)} \bigvee_{\boldsymbol{z}(k_e)} \cdots \bigvee_{\boldsymbol{z}(0):Poss(\boldsymbol{z}(0),\boldsymbol{f})=1} \tag{7.26}$$

$$\bigwedge_{k=0}^{k_e} \left(\bigvee_{\boldsymbol{r}(k)\in\mathcal{N}_{\boldsymbol{r}}} \bigwedge_{i=1}^{\nu} \chi_{\boldsymbol{L}_i^n}(z_i(k+1), \boldsymbol{w}_i(k), \boldsymbol{r}_i(k), z_i(k), \boldsymbol{v}_i(k), \boldsymbol{k}_i^T \cdot \boldsymbol{r}(k), f_i)\right)$$

has to be proven for each $0 \le k \le k_e$ to prove the theorem in the general case. This proof is held by induction.

Induction basis. For $k_e = 0$, the initial condition $Poss(z_i(0), f_i) = 1\ \forall (z_i(0), f_i) \in \mathcal{Z}_{0i} \times \mathcal{N}_{f_i}$ is inserted into eqn. (7.3). The result is entered into the left-hand side of eqn. (7.26) to obtain

$$\bigvee_{\boldsymbol{r}(0)} \bigwedge_{i=1}^{\nu} Poss(\boldsymbol{k}_i^T \cdot \boldsymbol{r}_i(k), \boldsymbol{r}_i(k), f_i, \boldsymbol{v}_i(0), \boldsymbol{w}_i(0)) \wedge Poss(\boldsymbol{f}, -1) \tag{7.27}$$

$$= \bigvee_{\boldsymbol{r}(0)} \bigwedge_{i=1}^{\nu} \left(\bigvee_{z_i(1)} \bigvee_{z_i(0):Poss(z_i(0),f_i)=1} \chi_{\boldsymbol{L}_i^n}(z_i(1), \boldsymbol{w}_i(0), \boldsymbol{r}_i(0), z_i(0), \boldsymbol{v}_i(0), \boldsymbol{s}_i(0), f_i) \right)$$
$$\wedge\, Poss(\boldsymbol{f}, -1)$$
$$= \bigvee_{\boldsymbol{z}(1)} \bigvee_{\boldsymbol{z}(0):\bigwedge_{i=1}^{\nu} Poss(z_i(0),f_i)=1} \left(\bigvee_{\boldsymbol{r}(0)} \bigwedge_{i=1}^{\nu} \chi_{\boldsymbol{L}_i^n}(z_i(1), \boldsymbol{w}_i(0), \boldsymbol{r}_i(0), z_i(0), \boldsymbol{v}_i(0), \boldsymbol{s}_i(0), f_i) \right)$$
$$\wedge\, Poss(\boldsymbol{f}, -1)$$
$$= \bigvee_{\boldsymbol{z}(1)} \bigvee_{\boldsymbol{z}(0):Poss(\boldsymbol{z}(0),\boldsymbol{f}))=1} \left(\bigvee_{\boldsymbol{r}(0)} \bigwedge_{i=1}^{\nu} \chi_{\boldsymbol{L}_i^n}(z_i(1), \boldsymbol{w}_i(0), \boldsymbol{r}_i(0), z_i(0), \boldsymbol{v}_i(0), \boldsymbol{s}_i(0), f_i) \right)$$

which is equal to the right-hand side of eqn. (7.26) for $k_e = 0$ and proves the induction basis for the general case.

Inductive step. The inductive hypothesis is proposed by assuming that eqn. (7.26) holds for some $k_e = k_h$ and is satisfied for $k_e = k_h + 1$ based on this hypothesis

$$\bigvee_{\boldsymbol{r}(k_h+1)} \bigwedge_{i=1}^{\nu} Poss(\boldsymbol{k}_i^T \cdot \boldsymbol{r}_i(k_h+1), \boldsymbol{r}_i(k_h+1), f_i, \boldsymbol{V}_i(0\ldots k_h+1), \boldsymbol{W}_i(0\ldots k_h+1)) \tag{7.28}$$
$$\wedge\, Poss(\boldsymbol{f}, k_h)$$
$$= \bigvee_{\boldsymbol{z}(k_h+2)} \bigvee_{\boldsymbol{z}(k_h+1)} \bigvee_{\boldsymbol{z}(k_h)} \cdots \bigvee_{\boldsymbol{z}(0):Poss(\boldsymbol{z}(0),\boldsymbol{f})=1} \bigwedge_{k=0}^{k_h+1}$$
$$\left(\bigvee_{\boldsymbol{r}(k)} \bigwedge_{i=1}^{\nu} \chi_{\boldsymbol{L}_i^n}(z_i(k+1), \boldsymbol{w}_i(k), \boldsymbol{r}_i(k), z_i(k), \boldsymbol{v}_i(k), \boldsymbol{k}_i^T \cdot \boldsymbol{r}(k), f_i) \right).$$

The right-hand side of this equation can be reformulated as

$$\bigvee_{\boldsymbol{z}(k_h+2)} \bigvee_{\boldsymbol{z}(k_h+1)} \bigvee_{\boldsymbol{z}(k_h)} \cdots \bigvee_{\boldsymbol{z}(0):Poss(\boldsymbol{z}(0),\boldsymbol{f})=1} \bigwedge_{k=0}^{k_h+1}$$
$$\left(\bigvee_{\boldsymbol{r}(k)} \bigwedge_{i=1}^{\nu} \chi_{\boldsymbol{L}_i^n}(z_i(k+1), \boldsymbol{w}_i(k), \boldsymbol{r}_i(k), z_i(k), \boldsymbol{v}_i(k), \boldsymbol{k}_i^T \cdot \boldsymbol{r}(k), f_i) \right)$$
$$= \bigvee_{\boldsymbol{z}(k_h+2)} \left(\bigvee_{\boldsymbol{r}(k_h+1)} \bigwedge_{i=1}^{\nu} \right. \tag{7.29}$$
$$\left. \chi_{\boldsymbol{L}_i^n}(z_i(k_h+2), \boldsymbol{w}_i(k_h+1), \boldsymbol{r}_i(k_h+1), z_i(k_h+1), \boldsymbol{v}_i(k_h+1), \boldsymbol{k}_i^T \cdot \boldsymbol{r}(k_h+1), f_i) \right)$$
$$\wedge \bigvee_{\boldsymbol{z}(k_h+1)} \bigvee_{\boldsymbol{z}(k_h)} \cdots \bigvee_{\boldsymbol{z}(0):Poss(\boldsymbol{z}(0),\boldsymbol{f})=1} \bigwedge_{k=0}^{k_h}$$

$$\left(\bigvee_{\boldsymbol{r}(k)} \bigwedge_{i=1}^{\nu} \chi_{\boldsymbol{L}_i^n}(z_i(k+1), \boldsymbol{w}_i(k), \boldsymbol{r}_i(k), z_i(k), \boldsymbol{v}_i(k), \boldsymbol{k}_i^T \cdot \boldsymbol{r}(k), f_i)\right)$$

$$= \bigvee_{\boldsymbol{z}(k_h+2)} \left(\bigvee_{\boldsymbol{r}(k_h+1)} \bigwedge_{i=1}^{\nu} \right. \tag{7.30}$$

$$\left. \chi_{\boldsymbol{L}_i^n}(z_i(k_h+2), \boldsymbol{w}_i(k_h+1), \boldsymbol{r}_i(k_h+1), z_i(k_h+1), \boldsymbol{v}_i(k_h+1), \boldsymbol{k}_i^T \cdot \boldsymbol{r}(k_h+1), f_i)\right)$$

$$\wedge \bigvee_{\boldsymbol{z}(k_h+1)} \bigvee_{\boldsymbol{z}(k_h)} \cdots \bigvee_{\boldsymbol{z}(0): \bigwedge_{i=1}^{\nu} Poss(z_i(0), f_i)=1} \bigwedge_{k=0}^{k_h}$$

$$\left(\bigvee_{\boldsymbol{r}(k)} \bigwedge_{i=1}^{\nu} \chi_{\boldsymbol{L}_i^n}(z_i(k+1), \boldsymbol{w}_i(k), \boldsymbol{r}_i(k), z_i(k), \boldsymbol{v}_i(k), \boldsymbol{k}_i^T \cdot \boldsymbol{r}(k), f_i)\right) \wedge Poss(\boldsymbol{f}, k_h)$$

$$= \bigwedge_{i=1}^{\nu} \bigvee_{z_i(k_h+2)} \bigvee_{\boldsymbol{r}_i(k_h+1)} \tag{7.31}$$

$$\chi_{\boldsymbol{L}_i^n}(z_i(k_h+2), \boldsymbol{w}_i(k_h+1), \boldsymbol{r}_i(k_h+1), z_i(k_h+1), \boldsymbol{v}_i(k_h+1), \boldsymbol{k}_i^T \cdot \boldsymbol{r}(k_h+1), f_i)$$

$$\wedge \left(\bigvee_{\boldsymbol{r}(k_h)} \bigwedge_{i=1}^{\nu} Poss(\boldsymbol{k}_i^T \cdot \boldsymbol{r}_i(k_h), \boldsymbol{r}_i(k_h), f_i, \boldsymbol{V}_i(0\dots k_h), \boldsymbol{W}_i(0\dots k_h))\right) \wedge Poss(\boldsymbol{f}, k_h)$$

$$= \bigvee_{\boldsymbol{r}(k_h+1)} \bigwedge_{i=1}^{\nu} Poss(\boldsymbol{k}_i^T \cdot \boldsymbol{r}_i(k_h+1), \boldsymbol{r}_i(k_h+1), f_i, \boldsymbol{V}_i(0\dots k_h+1), \boldsymbol{W}_i(0\dots k_h+1))$$

$$\wedge\, Poss(\boldsymbol{f}, k_h) \tag{7.32}$$

which proves the inductive step for the general case and the theorem. ■

7.3 Complexity considerations

The complexity of the partially coordinated diagnostic algorithm is investigated in this section in analogy to complexity considerations stated in Section 6.4 for the decentralized set-up.

Memory consumption. The maximal number of transitions that need to be stored for the network is given by eqn. (6.37). It cannot be further reduced as done in the decentralized set-up because the interconnection signals have to be taken into account in Algorithm 9 to estimate their possible values.

The memory consumption of the coordinator is very low because it has to store only the coupling model $\boldsymbol{K}$ which is a $\left(\sum_{i=1}^{\nu} \pi_i\right) \times \left(\sum_{i=1}^{\nu} \tau_i\right)$-matrix. Thus, it is negligible compared to the storage capacity of the component models.

Computational costs. The computational costs of the local diagnostic algorithm in partially coordinated diagnosis are higher than that of decentralized diagnosis because the interconnection signals must to be considered. The difference in the amount of operations between state-dependent autonomy and the general case is negligible compared to the total amount because it varies in the evaluation of eqn. (7.7). The majority of arithmetic operations is given by the prediction and the projection step. It amounts to[1]

$$N_i \cdot T_i \cdot P_i \cdot N_i \cdot S_i + P_i \cdot T_i \cdot S_i \cdot N_i \cdot N_i \tag{7.33}$$

where the first summand corresponds to the prediction step and the second one to the projection step. The number of operations of this projection step is, compared to that one used in the local diagnostic algorithm in the decentralized approach without preprocessing, N_i-times higher because eqn. (7.3) cannot use the results of the current prediction step. The order of eqn. (7.33) is $O(N_i^2 \cdot S_i \cdot T_i \cdot P_i)$ as opposed to eqn. (6.42) which is of order $O(N_i^2 \cdot S_i)$. Even though the dominating argument is N_i^2 for both terms, the influence of the coupling signals on the computational costs cannot be neglected because the relations $P_i = \prod_{j=1}^{\pi_i} P_i^j$ and $T_i = \prod_{j=1}^{\tau_i} T_i^j$ hold. Consequently, the component models should be as small as possible to ensure computability.

The dominating part of the computational costs of the coordinator is given by calculating the possible faults $\boldsymbol{f}$. In state-dependent autonomy, eqn. (7.8) is used. It requires the evaluation of

$$\nu \cdot S \tag{7.34}$$

arithmetic operations where $S = \prod_{i=1}^{\nu} S_i$ holds. Equation (7.13) necessitates

$$\nu \cdot T \cdot S \tag{7.35}$$

calculations in the general case with $T = \prod_{i=1}^{\nu} T_i$. Thus, it is of the order $O(\nu \cdot T \cdot S)$ which is T-times higher than the order $O(\nu \cdot S)$ in the case of autonomy. This difference is arises from the need of refining the locally obtained sets of possible pairs by applying the reduced composition rule. The corresponding search for fixed-points necessitates the evaluation of eqn. (7.13) only for all interconnection outputs due to their dependence with the interconnection inputs. The number of possible pairs $(\boldsymbol{s}_i(k_e), \boldsymbol{r}_i(k_e), f_i) \in \mathcal{R}_i(k_e)$ obtained by a local diagnostic unit is typically much smaller than $P_i \cdot T_i \cdot S_i$ in the worst case such that the computational costs are in general much lower than given by eqns. (7.34) and (7.35) but remain higher than in state-dependent autonomy. Consequently, the computational costs are in the case of autonomy a little lower in the local diagnostic units and significantly lower in the coordinator.

[1]The index "i" is used in this section to explicitly indicate the correspondence with the component models.

7.4 Comparative example

This section gives an example to compare the different diagnostic approaches presented in this thesis for the general case and the simplifications resulting from state-dependent autonomy. The system under consideration is shown in Fig. 7.2.

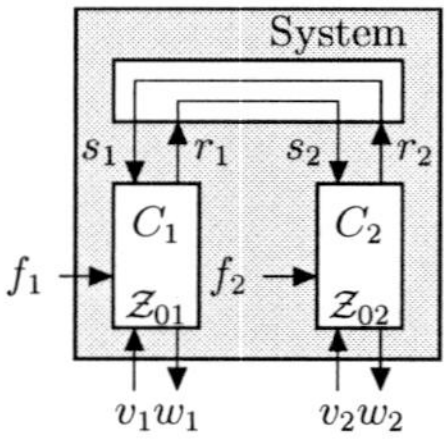

Figure 7.2: Interconnected discrete-event system

It consists of two interacting components C_1 and C_2 whose behaviors are given by the automata graphs depicted in Fig. 7.3(a) and 7.3(b) respectively. The composition of the network yields the I/O-automaton $\tilde{\mathcal{N}}$ with the automaton graph shown in Fig. 7.4. The asterisk $*$ denotes any value of the fault set.

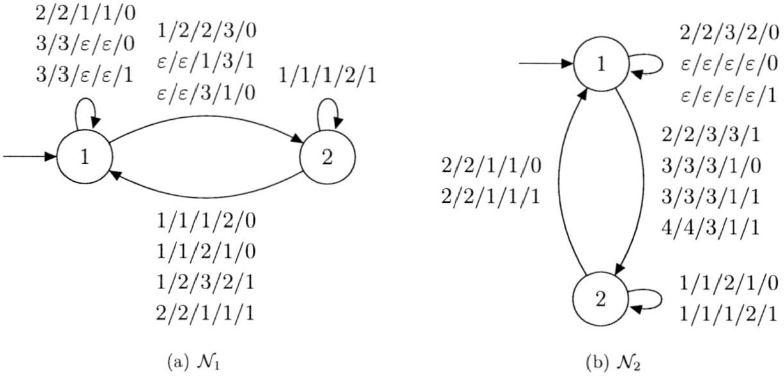

Figure 7.3: Component models

It is assumed that the following measurements are given:

$$\tilde{B}_1(4) = ((3/3), (\varepsilon/\varepsilon), (1/1), (2/2), (1/2))$$
$$\tilde{B}_2(4) = ((\varepsilon/\varepsilon), (3/3), (1/1), (2/2), (2/2)).$$

The diagnostic results obtained by the centralized Algorithm 6 are shown in Table 7.1. The results of the decentralized diagnostic Algorithms 7 and 8 are summarized in Table 7.2. Finally, partially coordinated diagnosis is considered in Table 7.3.

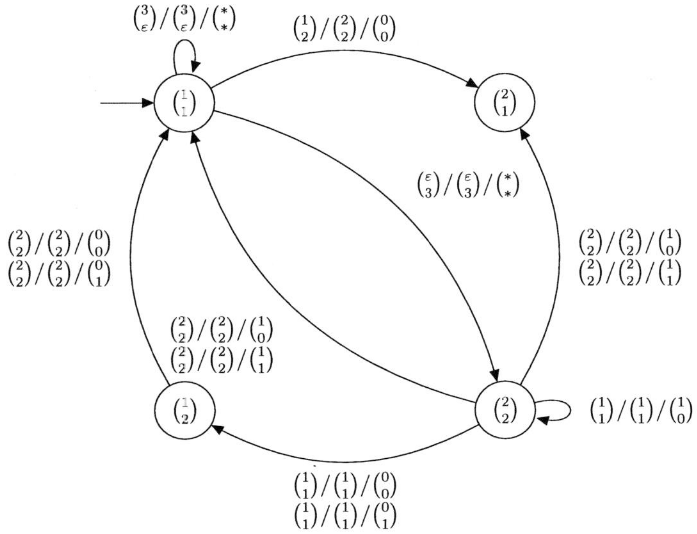

Figure 7.4: Composed I/O-automaton $\tilde{\mathcal{N}}$

Table 7.1: Centralized diagnosis using the I/O-automaton $\tilde{\mathcal{N}}$

	$k_e = -1$	$k_e = 0$	$k_e = 1$	$k_e = 2$	$k_e = 3$	$k_e = 4$
Prediction $\{(\boldsymbol{z}, \boldsymbol{f})\}$	$\{(\binom{1}{1}, \binom{0}{0}), (\binom{1}{1}, \binom{1}{0}), (\binom{1}{1}, \binom{0}{1}), (\binom{1}{1}, \binom{1}{1})\}$	$\{(\binom{1}{1}, \binom{0}{0}), (\binom{1}{1}, \binom{1}{0}), (\binom{1}{1}, \binom{0}{1}), (\binom{1}{1}, \binom{1}{1})\}$	$\{(\binom{2}{2}, \binom{0}{0}), (\binom{2}{2}, \binom{1}{0}), (\binom{2}{2}, \binom{0}{1}), (\binom{2}{2}, \binom{1}{1})\}$	$\{(\binom{1}{2}, \binom{0}{0}), (\binom{1}{2}, \binom{0}{1}), (\binom{2}{2}, \binom{1}{0})\}$	$\{(\binom{1}{1}, \binom{0}{0}), (\binom{1}{1}, \binom{0}{1}), (\binom{2}{1}, \binom{1}{0})\}$	$\{(\binom{2}{1}, \binom{0}{0})\}$
Projection $\mathcal{F}^c(k_e) = \{\boldsymbol{f}\}$	–	$\{\binom{0}{0}, \binom{1}{0}, \binom{0}{1}, \binom{1}{1}\}$	$\{\binom{0}{0}, \binom{1}{0}, \binom{0}{1}, \binom{1}{1}\}$	$\{\binom{0}{0}, \binom{1}{0}, \binom{0}{1}\}$	$\{\binom{0}{0}, \binom{1}{0}, \binom{0}{1}\}$	$\{\binom{0}{0}\}$

Time $k_e = 0$. The measurements $(3/3)$ and $(\varepsilon/\varepsilon)$ correspond to the asynchronous motion of component C_1 because of the missing interaction with the component C_2 due to $s_1 = \varepsilon = r_1$. The same diagnostic results are obtained by all three approaches because both components fulfill the autonomy condition (6.32) in addition.

Time $k_e = 1$. The measurements $(\varepsilon/\varepsilon)$ and $(3/3)$ result in a synchronous motion of both components where C_2 causes the state transition of C_1 which gets a vanishing input symbol. The autonomy condition (6.32) holds like in the previous step such that all algorithms yield the same diagnostic result.

Table 7.2: Decentralized diagnosis using the I/O-automata $\mathcal{N}_1$ and $\mathcal{N}_2$

	$k_e=-1$	$k_e=0$	$k_e=1$	$k_e=2$	$k_e=3$	$k_e=4$
Prediction 1 $\{(z_1, f_1)\}$	$\{(1,0), (1,1)\}$	$\{(1,0), (1,1)\}$	$\{(2,0), (2,1)\}$	$\{(1,0), (2,1)\}$	$\{(1,0), (2,1)\}$	$\{(2,0), (1,1)\}$
Prediction 2 $\{(z_2, f_2)\}$	$\{(1,0), (1,1)\}$	$\{(1,0), (1,1)\}$	$\{(2,0), (2,1)\}$	$\{(2,0), (2,1)\}$	$\{(1,0), (1,1)\}$	$\{(1,0), (2,1)\}$
Projection 1 $\mathcal{F}_1^{dc}(k_e)=\{f_1\}$	–	$\{0,1\}$	$\{0,1\}$	$\{0,1\}$	$\{0,1\}$	$\{0,1\}$
Projection 2 $\mathcal{F}_2^{dc}(k_e)=\{f_2\}$	–	$\{0,1\}$	$\{0,1\}$	$\{0,1\}$	$\{0,1\}$	$\{0,1\}$
$\mathcal{F}^{dc}(k_e)$	–	$\{\binom{0}{0}, \binom{1}{0}, \binom{0}{1}, \binom{1}{1}\}$	$\{\binom{0}{0}, \binom{1}{0}, \binom{0}{1}, \binom{1}{1}\}$	$\{\binom{0}{0}, \binom{1}{0}, \binom{0}{1}, \binom{1}{1}\}$	$\{\binom{0}{0}, \binom{1}{0}, \binom{0}{1}, \binom{1}{1}\}$	$\{\binom{0}{0}, \binom{1}{0}, \binom{0}{1}, \binom{1}{1}\}$

Time $k_e = 2$. The autonomy condition (6.32) is violated for the first I/O-automaton because of the different successor states for the measurements (1/1) but holds for the second I/O-automaton for the measurements (1/1). The coordination of the locally obtained sets of possible pairs does not result in any improvements of the result but restricts the possible combinations of locally obtained faults from $\{0,1\} \times \{0,1\}$ to $\{\binom{0}{0}, \binom{1}{0}, \binom{0}{1}\}$, i.e. the fault $\binom{1}{1}$ has been rejected from the partially coordinated set of possible faults. The same result has been obtained by the centralized diagnostic approach whereas the decentralized set-up yields all possible combinations of locally obtained faults.

Time $k_e = 3$. The need to store the partially coordinated set of possible faults in the coordinator for the use in the next time step is clarified for the measurements (2/2) and (2/2). The coordination of the locally obtained sets of possible pairs allows for all combinations of $\{0,1\} \times \{0,1\}$ to be possible faults which results in an increased partially coordinated set of possible faults. The only way to maintain the restrictions on the possible faults obtained in the previous step is to store them in the coordinator.

Time $k_e = 4$. The restrictions among the coupling signals reduce the centrally obtained and partially coordinated sets of possible faults for the measurements (1/2) and (2/2) further compared to the decentrally obtained result because only the combination $\binom{2}{3}/\binom{3}{2}$ results in a fixed point that solves the feedback problem. Correspondingly, the faults $\{\binom{1}{0}, \binom{0}{1}\}$ can be rejected from the partially coordinated set of possible faults. As a result, the system is considered to operate fault free by centralized and partially coordinated diagnosis based on the measurements whereas the decentrally obtained set of possible faults still consists of all possible combinations of faults $\{0,1\} \times \{0,1\}$.

Table 7.3: Partially coordinated diagnosis using the I/O-automata $\mathcal{N}_1$ and $\mathcal{N}_2$

	$k_e=-1$	$k_e=0$	$k_e=1$	$k_e=2$	$k_e=3$	$k_e=4$
Projection 1 $\mathcal{R}_1(k_e)=\{(s_1,r_1,f_1)\}$	–	$\{(\varepsilon,\varepsilon,0),(\varepsilon,\varepsilon,1)\}$	$\{(1,3,0),(3,1,1)\}$	$\{(1,2,0),(2,1,0),(1,2,1)\}$	$\{(1,1,0),(1,1,1)\}$	$\{(2,3,0),(3,2,1)\}$
Projection 2 $\mathcal{R}_2(k_e)=\{(s_2,r_2,f_2)\}$	–	$\{(\varepsilon,\varepsilon,0),(\varepsilon,\varepsilon,1)\}$	$\{(3,1,0),(3,1,1),(1,3,0),(1,3,1)\}$	$\{(2,1,0),(1,2,1)\}$	$\{(1,1,0),(1,1,1)\}$	$\{(3,2,0),(3,3,1)\}$
Coordination $\mathcal{J}(k_e)=\{(\boldsymbol{s},\boldsymbol{r},\boldsymbol{f})\}$	–	$\{(\binom{\varepsilon}{\varepsilon},\binom{\varepsilon}{\varepsilon},\binom{0}{0}),(\binom{\varepsilon}{\varepsilon},\binom{\varepsilon}{\varepsilon},\binom{1}{0}),(\binom{\varepsilon}{\varepsilon},\binom{\varepsilon}{\varepsilon},\binom{0}{1}),(\binom{\varepsilon}{\varepsilon},\binom{\varepsilon}{\varepsilon},\binom{1}{1})\}$	$\{(\binom{1}{3},\binom{3}{1},\binom{0}{0}),(\binom{1}{3},\binom{3}{1},\binom{0}{1}),(\binom{3}{1},\binom{1}{3},\binom{1}{0}),(\binom{3}{1},\binom{1}{3},\binom{1}{1})\}$	$\{(\binom{1}{2},\binom{2}{1},\binom{0}{0}),(\binom{1}{2},\binom{2}{1},\binom{1}{0}),(\binom{2}{1},\binom{1}{2},\binom{0}{1})\}$	$\{(\binom{1}{1},\binom{1}{1},\binom{0}{0}),(\binom{1}{1},\binom{1}{1},\binom{1}{0}),(\binom{1}{1},\binom{1}{1},\binom{0}{1}),(\binom{1}{1},\binom{1}{1},\binom{1}{1})\}$	$\{(\binom{2}{3},\binom{3}{2},\binom{0}{0})\}$
$\mathcal{F}^{pc}(k_e)$	–	$\{\binom{0}{0},\binom{1}{0},\binom{0}{1},\binom{1}{1}\}$	$\{\binom{0}{0},\binom{1}{0},\binom{0}{1},\binom{1}{1}\}$	$\{\binom{0}{0},\binom{1}{0},\binom{0}{1}\}$	$\{\binom{0}{0},\binom{1}{0},\binom{0}{1}\}$	$\{\binom{0}{0}\}$
Improvement 1 $\mathcal{R}_1'(k_e)=\{(s_1,r_1,f_1)\}$	–	$\emptyset$	$\emptyset$	$\emptyset$	$\emptyset$	$\{(3,2,1)\}$
Improvement 2 $\mathcal{R}_2'(k_e)=\{(s_2,r_2,f_2)\}$	–	$\emptyset$	$\emptyset$	$\emptyset$	$\emptyset$	$\{(3,3,1)\}$
Prediction 1 $\{(z_1,f_1)\}$	$\{(1,0),(1,1)\}$	$\{(1,0),(1,1)\}$	$\{(2,0),(2,1)\}$	$\{(1,0),(2,1)\}$	$\{(1,0),(2,1)\}$	$\{(2,0)\}$
Prediction 2 $\{(z_2,f_2)\}$	$\{(1,0),(1,1)\}$	$\{(1,0),(1,1)\}$	$\{(2,0),(2,1)\}$	$\{(2,0),(2,1)\}$	$\{(1,0),(1,1)\}$	$\{(1,0)\}$

7.5 Evaluation of the presented diagnostic approaches

Three different information structures have been investigated in this thesis for the diagnosis of coupled mechatronical systems with asynchronous behavior. The centralized approach has been proven to yield the ideal diagnostic result but does not reduce the complexity. This drawback can be partially overcome by using online composition. As composition accompanies the loss of network effects like state-dependent autonomy and asynchronous motion of several components, the presented algorithms do not differ from the diagnostic algorithms for synchronous networks of I/O-automata considered in [84].

The remaining inflexibility of the centralized set-up is solved by splitting the diagnostic task into several smaller subtasks. The locally obtained diagnostic results are combined in the merger to the decentrally obtained set of possible faults. The asynchronous behavior of components is indicated in this thesis by the explicit use of the empty symbol ε to ensure equality of the length of all sequences of local measurements. Consequently, the locally obtained diagnostic results are synchronized upon the occurrence of events as opposed to

[84] where a clock signal is used for synchronization purposes. This decentrally obtained diagnostic result is proven to be complete and sound in the case of state-dependent autonomy. A lack of soundness arises in the general case from the neglected dependencies in the local calculations between the components but it is still guaranteed that no fault candidate is excluded wrongly.

The ideal diagnostic result can only be obtained by estimating the values of the interconnection signals locally and removing contradictions between these values in the centralized coordinator. The drawbacks of partially coordinated diagnosis are a higher amount of needed memory and higher computational costs in the local diagnostic units because of regarding the coupling signals. It has been shown that the storage capacity and the number of arithmetic operations evaluated in the coordinator are in general higher than in the decentralized set-up but they are equal in both approaches in state-dependent autonomy. These results are summarized in Table 7.4.

Table 7.4: Comparison of the diagnostic approaches

	Centralized	**Decentralized**		**Partially coordinated**	
		autonomy	general case	autonomy	general case
Local computations		$O(N_i^2 S_i)$, $i \in \{1, \ldots, \nu\}$		$O(N_i^2 S_i P_i T_i)$, $i \in \{1, \ldots, \nu\}$	
Central computations	$O(N^2 S)$	$O(\nu S)$		$O(\nu S)$	$O(\nu TS)$
Local memory		$O(N_i^2 M_i R_i S_i)$, $i \in \{1, \ldots, \nu\}$		$O(N_i^2 M_i R_i P_i T_i S_i)$, $i \in \{1, \ldots, \nu\}$	
Central memory	$O((N^2 MRS)^\nu)$	$O(\nu S)$		$O(\nu S)$	$O(\nu TS)$
Quality of the results	+	+	−	+	+
Completeness	+	+	+	+	+
Soundness	+	+	−	+	+
Scalability	poor	very high		high	
Reusability	poor	high		high	
Reliability	poor	high		high	

The algorithms for partially coordinated diagnosis developed in this thesis differ in three ways from that algorithms presented in [84]:

1. The formalism is adapted to handle models that are based on behavioral relations instead of relational algebra.

2. The amount of exchanged data sent from the coordinator to the local diagnostic units is reduced in the case of state-dependent autonomy.

3. It is shown that the partially coordinated diagnostic result must be stored in the coordinator for a correct calculation of the partially coordinated set of possible faults in the next time step. Hence, the coordinator must have a memory which is not proposed in [84] wrongly.

As all sequences of local measurements are of equal length due to the use of the empty symbol ε, the algorithms developed in this thesis can be applied directly to synchronous networks of I/O-automata. The only difference is the way of updating the signals of the network: A clock signal is used in the synchronous case as opposed to the occurrence of events in the asynchronous case.

Chapter 8

Application examples

The chapter gives two examples to demonstrate the application of the approaches presented in this thesis. Decentralized and partially coordinated diagnoses of a process plant are studied in Section 8.1. The simulation of a production facility using online composition is considered in Section 8.2.

8.1 Partially coordinated diagnosis of a process plant

8.1.1 Modeling

The application of the partially coordinated diagnostic Algorithms 9 and 10 to the process plant shown in Fig. 8.1 is investigated in this section. It relies on the Study thesis [9]. The vessel has discrete senors to measure the filling level and the temperature of the contained liquid. They have the values "empty" and "full" respectively "cold" and "hot". The in- and

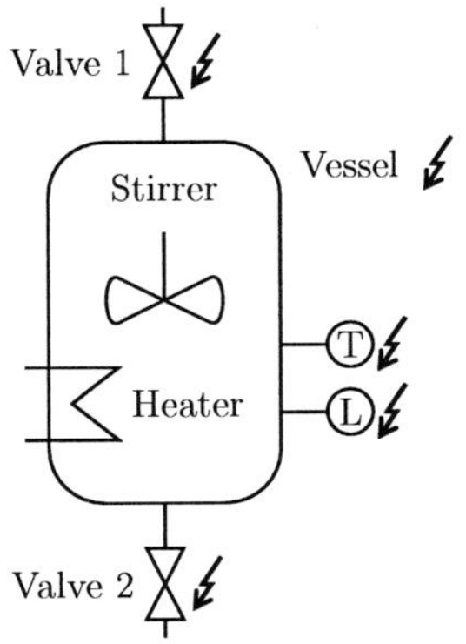

Figure 8.1: The process plant from [9]

outflow is controlled by valves 1 and 2. The heater is used to regulate the temperature. Its use is only allowed, if the stirrer is turned on. The system is assumed to be subject to the following faults:

- Valve 1 respectively 2 blocks
- Breakdown of the compressed air at valve 1 respectively 2
- Leakage at the bottom of the vessel
- Breakdown of the level sensor
- Breakdown of the temperature sensor.

Each component of the plant is modeled as a nondeterministic I/O-automaton given by eqn. (3.29). The interactions between these components are shown in Fig. 8.2. The resulting interconnection structure of the model using the coupling model to link the interconnection signals is given in Fig. 8.3.

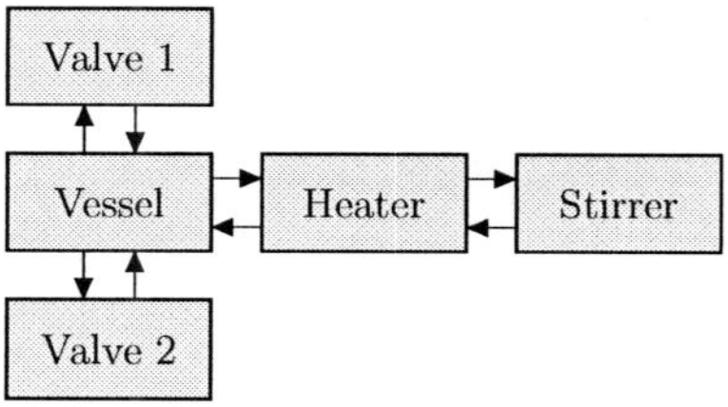

Figure 8.2: Structure of the process plant [9]

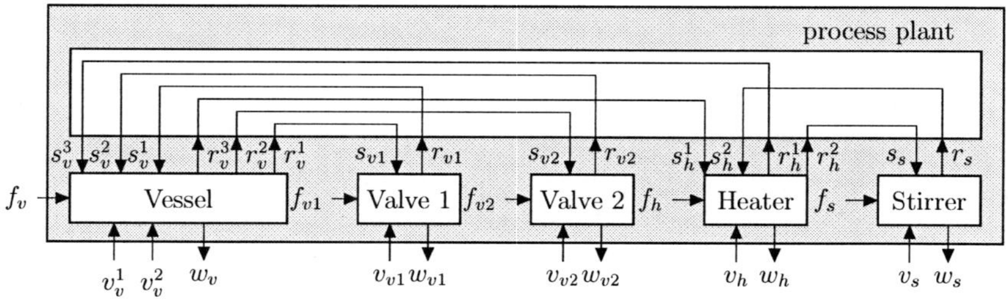

Figure 8.3: Interconnection structure of the model

Vessel model. The vessel can be seen in this configuration as a master unit. It is modeled with four states. It receives the measurements of the level sensor and of the temperature sensor as control inputs and outputs its state. The coupling signals s_v^1 and r_v^1 describe the interaction with valve 1 whereas s_v^2 and r_v^2 are used for the interaction with valve 2. The communication with the heater is realized by s_v^3 and r_v^3. In fault-free operation, the filling level can increase, only if valve 2 is opened and valve 1 is closed. The filling level is reduced, if valve 2 is closed and valve 1 is opened. The filling level degrades independently from the position of the valve, if the fault "Leakage at the bottom of the vessel" occurs. The temperature rises, only if the heater is turned on and falls, if it is turned off. The level sensor is broken, if either the temperature falls while the heater is turned on or the temperature rises while the heater is turned off.

The sets of states and faults are given in Appendix B.1.1 in Table B.1, the control signals in Table B.2, the sets of interconnection inputs in Table B.3 and the set of interconnection outputs in Table B.4. The behavioral relation of the vessel subject to fault $f_v = 0$ is shown in Table B.5, subject to fault $f_v = 1$ in Table B.6, subject to fault $f_v = 2$ in Table B.7 and subject to fault $f_v = 3$ in Table B.8.

Valve model. Valves 1 and 2 are described by the same model. They can be controlled via their control inputs. They can change their states with or without communicating it to the vessel via the interconnection signals. The vessel can request the current state of the valves.

The sets of states and faults are given in Appendix B.1.2 in Table B.9, the control signals in Table B.10, the sets of interconnection signals in Table B.11 and the set of interconnection outputs in Table B.32. The behavioral relation of the fault-free valves are given in Table B.12 and subject to the considered faults in Table B.13.

Heater model. The heater can be turned on or off externally. Its state is checked in conjunction with state changes of the vessel via the interconnection signals s_h^1 and r_h^1. The heater checks itself the state of the stirrer via the interconnection signals s_h^2 and r_h^2 because it cannot be turned on, if the stirrer is not operating.

The sets of states and faults are given in Appendix B.1.3 in Table B.14, the control signals in Table B.15, the sets of interconnection signals in Table B.16 and the set of interconnection outputs in Table B.32. The behavioral relation of the heater is given in Table B.17.

Stirrer model. The stirrer is controlled externally. It turns the heater via its interconnection output off, if it is turned off.

The sets of states and faults are given in Appendix B.1.4 in Table B.18, the control signals in Table B.19, the sets of interconnection signals in Table B.20 and the set of interconnection outputs in Table B.32. The behavioral relation of the stirrer is given in Table B.21.

8.1.2 Diagnosis

The decentralized and the partially coordinated diagnostic algorithm are applied in this section to the process plant shown in Fig. 8.1. It is assumed that the sequences of control inputs and outputs given in Table 8.1 have been measured. The obtained diagnostic results are discussed for each time step regarding the simplifications resulting from state-dependent autonomy.

Table 8.1: Measured sequence of control inputs and outputs of the process plant

	k_e	0	1	2	3
Vessel	v_v^1	ε	2	ε	1
	v_v^2	ε	ε	ε	ε
	w_v	ε	4	ε	2
Valve 1	v_{v1}	2	ε	ε	ε
	w_{v1}	2	ε	ε	ε
Valve 2	v_{v2}	1	ε	ε	ε
	w_{v2}	1	ε	ε	ε
Heater	v_h	ε	ε	ε	ε
	w_h	ε	ε	ε	ε
Stirrer	v_s	ε	ε	1	ε
	w_s	ε	ε	1	ε

Time $k_e = 0$. The local diagnostic unit of the vessel gets the measurements $(v_v^1(0) = \varepsilon, v_v^2(0) = \varepsilon, w_v(0) = \varepsilon)$ which correspond to an asynchronous state transition that fulfills the autonomy condition (7.4). It sends the set of possible pairs $\{(\varepsilon, \varepsilon, \varepsilon, \varepsilon, \varepsilon, \varepsilon, *)\}$ to the coordinator and outputs the set of possible states and faults $\{(1,*), (2,*), (3,*), (4,*)\}$ resulting from prediction. No fault can be rejected.

The local diagnostic unit of valve 1 gets the measurements $(v_{v1}(0) = 2, w_{v1}(0) = 2)$ which correspond to the control input "open valve" and the control output "valve is opened". The locally obtained set of possible pairs $\{(\varepsilon, \varepsilon, 0), (1, 2, 0), (\varepsilon, \varepsilon, 1), (1, 2, 1)\}$ is obtained by projection and sent to the coordinator. As the autonomy condition (7.4) holds, the prediction step can be applied without an answer of the coordinator to obtain the set of possible states and faults $\{(2, 0), (2, 1)\}$. Thus, the fault $f_{v1} = 2$ can be rejected from the locally obtained diagnostic result.

The local diagnostic unit of valve 2 gets the measurements $(v_{v2}(0) = 1, w_{v2}(0) = 1)$ which correspond to the control input "close valve" and the control output "valve is closed". The locally obtained set of possible pairs $\{(\varepsilon, \varepsilon, 0), (1, 1, 0), (\varepsilon, \varepsilon, 1), (1, 1, 1), (\varepsilon, \varepsilon, 2), (1, 1, 2)\}$ is obtained by projection and is sent to the coordinator. The autonomy condition (7.4) holds such that the prediction step can be applied without an answer of the coordinator to get the set of possible states and faults $\{(1, 0), (1, 1), (1, 2)\}$. Hence, no faults can be rejected.

The local diagnostic unit of the heater gets the measurements $(v_h(0) = \varepsilon, w_h(0) = \varepsilon)$ which results either in an asynchronous state transition or a state transition that is caused by the stirrer. As the locally obtained set of possible pairs is given by $\{(\varepsilon, \varepsilon, \varepsilon, \varepsilon, 0), (\varepsilon, 1, \varepsilon, 1, 0)\}$, the autonomy condition (7.4) holds and the prediction can be used to obtain the set of possible states and faults $\{(1, 0), (2, 0)\}$.

The local diagnostic unit of the stirrer gets the measurements $(v_s(0) = \varepsilon, w_s(0) = \varepsilon)$ which results either in an asynchronous state transition or a state transition that is caused by the heater. As the locally obtained set of possible pairs is given by $\{(\varepsilon, \varepsilon, 0), (1, 1, 0)\}$, the autonomy condition (7.4) holds and the prediction can be used to obtain the set $\{(1, 0), (2, 0)\}$.

The coordinator gets the locally obtained sets of possible pairs from all local diagnostic units which fulfill the autonomy condition (7.4). Hence, there is no need to coordinate. As a result, the decentrally obtained diagnostic result $\mathcal{F}^{dc}(0)$ is equal to the partially coordinated one $\mathcal{F}^{pc}(0)$ (Table 8.2).

Table 8.2: Diagnostic results at time $k_e = 0$

(a) $\mathcal{F}^{dc}(0)$

f_v	f_{v1}	f_{v2}	f_h	f_s
0	0	0	0	0
1	1	1		
2		2		
3				

(b) $\mathcal{F}^{pc}(0)$

f_v	f_{v1}	f_{v2}	f_h	f_s
0	0	0	0	0
1	1	1		
2		2		
3				

Time $k_e = 1$. The local diagnostic unit of the vessel gets the measurements $(v_v^1(1) = 2, v_v^2(1) = \varepsilon, w_v(1) = 4)$ which correspond to the control input "Vessel is full" and the control output "Vessel is full & hot". The fault 2 can be rejected. The autonomy condition (7.4) does not hold for the faults $0, 1, 3$. The locally obtained set of possible pairs $\{(2, 1, \varepsilon, 1, 1, \varepsilon, 0), (1, 2, \varepsilon, 1, 1, \varepsilon, 1), (2, 1, \varepsilon, 1, 1, \varepsilon, 3)\}$ is sent to the coordinator. The local diagnostic unit waits for an answer of the coordinator before it evaluates the prediction step.

The local diagnostic unit of valve 1 gets the measurements $(v_{v1}(1) = \varepsilon, w_{v1}(1) = \varepsilon)$ which correspond either to an asynchronous state transition or a state transition caused by the vessel. The locally obtained set of possible pairs $\{(\varepsilon, \varepsilon, 0), (1, 2, 0), (\varepsilon, \varepsilon, 1), (1, 2, 1)\}$ is obtained by projection and sent to the coordinator. As the autonomy condition (7.4) holds, the prediction step can be applied without an answer of the coordinator to obtain the set of possible states and faults $\{(2, 0), (2, 1)\}$. Thus, no fault can be rejected.

The local diagnostic unit of valve 2 also gets the measurements $(v_{v2}(1) = \varepsilon, w_{v2}(1) = \varepsilon)$. The locally obtained set of possible pairs $\{(\varepsilon, \varepsilon, 0), (1, 1, 0), (\varepsilon, \varepsilon, 1), (1, 1, 1)\}$ is sent to the coordinator. As the autonomy condition (7.4) holds, the prediction step can be applied without an answer of the coordinator to obtain the set of possible states and faults $\{(1,0),(1,1),(1,2)\}$. Thus, no fault can be rejected.

The local diagnostic unit of the heater gets the measurements $(v_h(1) = \varepsilon, w_h(1) = \varepsilon)$. Hence, the same results are obtained as in the previous step. The results obtained by the local diagnostic unit of the stirrer are also equal to the previous results because it gets the measurements $(v_s(1) = \varepsilon, w_s(1) = \varepsilon)$.

The coordinator gets from all local diagnostic units the locally obtained sets of possible pairs. There is a need to coordinate because the vessel does not fulfill the autonomy condition (7.4). The coordinator removes the tuple $(1,2,\varepsilon,1,1,\varepsilon,1)$ from the locally obtained set of possible values of the local diagnostic unit of the vessel because it is in contradiction to the other locally obtained sets of possible pairs. The fault $f_v = 1$ can be rejected by the coordinator. Hence, the decentrally obtained diagnostic result degrades compared to the partially coordinated one (Table 8.3).

The vessel is the only component that gets improvements from the coordinator because all other components are in autonomy. The evaluation of the prediction step yields the set of possible states and faults $\{(4,0),(4,3)\}$.

Table 8.3: Diagnostic results at time $k_e = 1$

(a) $\mathcal{F}^{dc}(1)$

f_v	f_{v1}	f_{v2}	f_h	f_s
0	0	0	0	0
2	1	1		
3		2		

(b) $\mathcal{F}^{pc}(1)$

f_v	f_{v1}	f_{v2}	f_h	f_s
0	0	0	0	0
3	1	1		
		2		

Time $k_e = 2$. The local diagnostic unit of the vessel gets the measurements $(v_v^1(2) = \varepsilon, v_v^2(2) = \varepsilon, w_v(2) = \varepsilon)$ which correspond to an asynchronous state transition that fulfills the autonomy condition (7.4). The locally obtained set of possible pairs $\{(\varepsilon,\varepsilon,\varepsilon,\varepsilon,\varepsilon,\varepsilon,0), (\varepsilon,\varepsilon,\varepsilon,\varepsilon,\varepsilon,\varepsilon,3)\}$ is sent to the coordinator. The set of possible states and faults $\{(4,0),(4,3)\}$ is the result of the prediction step. No fault can be rejected.

The local diagnostic unit of valve 1 gets the measurements $(v_{v1}(2) = \varepsilon, w_{v1}(2) = \varepsilon)$, the one of valve 2 gets $(v_{v2}(2) = \varepsilon, w_{v2}(2) = \varepsilon)$ and the one of the heater gets $(v_h(2) = \varepsilon, w_h(2) = \varepsilon)$. Hence, the same results are obtained in all local diagnostic units as in the previous step.

The local diagnostic unit of the stirrer gets the measurements $(v_s(2) = 1, w_s(2) = 1)$ which indicates that the stirrer has been turned off by its environment. Hence, the stirrer must turn of the heater internally, if it is turned on. The corresponding locally obtained set of possible pairs is given as $\{(1,1,0),(2,3,0)\}$. Autonomy is not at hand such that the local diagnostic unit cannot continue until the coordinator has sent the improvements.

As not all components are in autonomy, coordination takes place and removes the tuple $(2, 3, 0)$ from the locally obtained set of possible pairs sent from the local diagnostic unit of the stirrer. This result is used in the prediction step to get the set of possible states and faults $\{(1, 0)\}$ as possible combination of successor state and fault of the stirrer. The obtained diagnostic results are given in Table 8.4.

Table 8.4: Diagnostic results at time $k_e = 2$

(a) $\mathcal{F}^{dc}(2)$

f_v	f_{v1}	f_{v2}	f_h	f_s
0	0	0	0	0
2	1	1		
3		2		

(b) $\mathcal{F}^{pc}(2)$

f_v	f_{v1}	f_{v2}	f_h	f_s
0	0	0	0	0
3	1	1		
		2		

Time $k_e = 3$. The local diagnostic unit of the vessel gets the measurements $(v_v^1(3) = 1, v_v^2(3) = \varepsilon, w_v(3) = 1)$ which correspond to the control input "Vessel is empty" and the control outputs "Vessel is empty & hot". This state transition is allowed in the fault-free case only, if valve 1 is closed and valve 2 is opened whereas the opposite positions are allowed for fault $f_v = 3$. The locally obtained set of possible pairs $\{(1, 2, \varepsilon, 1, 1, \varepsilon, 0), (2, 1, \varepsilon, 1, 1, \varepsilon, 3), \}$ is sent to the coordinator. As autonomy is not given, the local diagnostic unit stops processing.

The local diagnostic unit of valve 1 gets the measurements $(v_{v1}(3) = \varepsilon, w_{v1}(3) = \varepsilon)$, that one of valve 2 gets $(v_{v2}(3) = \varepsilon, w_{v2}(3) = \varepsilon)$, that one of the heater gets $(v_h(3) = \varepsilon, w_h(3) = \varepsilon)$ and that one of the stirrer gets $(v_s(3) = \varepsilon, w_s(3) = \varepsilon)$. Hence, the same results are obtained in all local diagnostic units as in the previous step.

Coordination takes place because not all components are in autonomy. The tuple $(1, 2, \varepsilon, 1, 1, \varepsilon, 0)$, obtained by the local diagnostic unit of the vessel, is rejected because it is in contradiction to the results obtained by the local diagnostic units of valve 1 and 2. Hence, the fault $f_v = 0$ can be rejected by the partially coordinated diagnostic algorithm, whereas it is still possible in decentralized diagnosis (Table 8.5). The improvements are sent back to the local diagnostic units to update their results.

Table 8.5: Diagnostic results at time $k_e = 3$

(a) $\mathcal{F}^{dc}(3)$

f_v	f_{v1}	f_{v2}	f_h	f_s
0	0	0	0	0
2	1	1		
3		2		

(b) $\mathcal{F}^{pc}(3)$

f_v	f_{v1}	f_{v2}	f_h	f_s
3	0	0	0	0
	1	1		
		2		

8.2 Simulation of a production facility

8.2.1 Modeling

This section describes the application of the simulation Algorithm 3 to the small part of a production facility that is shown in Fig. 8.4 [70]. The presented results rely on the Diploma thesis [11]. The system consists of a control unit, a robot, two conveyers 1 and 2 and a leaving conveyer. The task of the robot is to take the work pieces A respectively B and place them in a given order on the leaving conveyer. The sequence of taking the work pieces is given by the sorting task which is stored in the control unit.

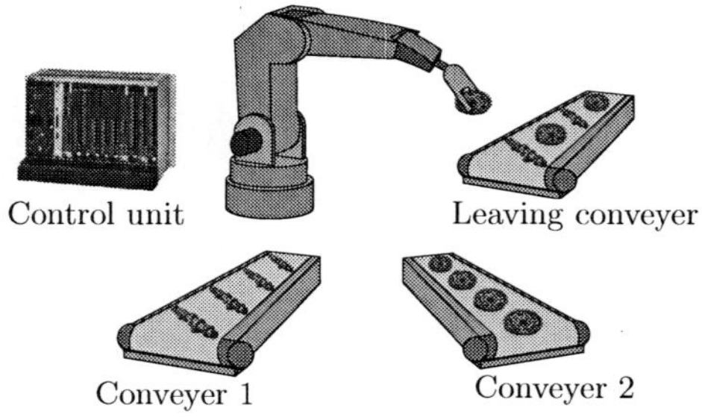

Figure 8.4: Sorting of work pieces by a robot [70]

The system is considered for modeling purposes as the set of five interacting components:

- The conveyer 1 delivers work piece A.
- The conveyer 2 delivers work piece B.
- The uncontrolled robot transports work pieces from one conveyer to another one.
- The leaving conveyer transports the sorted work piece to the next work station.
- The control unit operates the robot based on the desired sorting task.

Each component of the production facility is modeled as a deterministic I/O-automaton given by eqn. (3.11). The interactions between the components are depicted in Fig. 8.5. The resulting interconnection structure of the model using the coupling model link the interconnection signals is given in Fig. 3.7.

The control unit sends commands to each component via its interconnection outputs. Its interconnection inputs are used to inform the control unit about the states of the components. These commands are based on the following sorting task:

1. All components are activated externally.

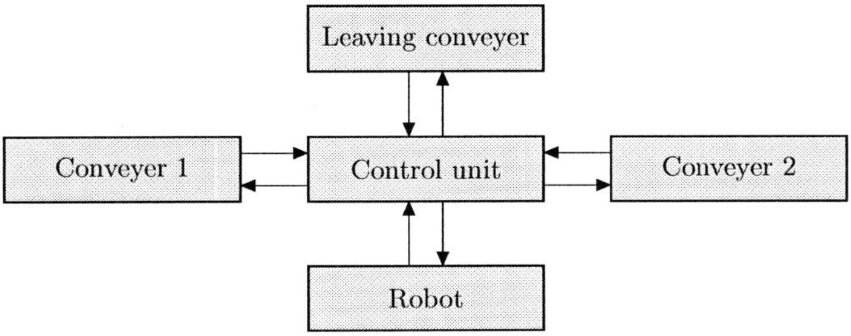

Figure 8.5: Structure of the production facility [11]

2. The control unit starts all conveyers and directs the robot to conveyer 1.
3. The robot takes one work piece A and moves to the leaving conveyer. A new work piece B is put at the same time on conveyer 2 externally. Conveyer 2 moves until the first work piece reaches its desired position such that is can be taken by the robot.
4. The robot deposits the taken work piece A on the leaving conveyer. It waits with an opened clamshell at that position for new instructions from the control unit. The leaving conveyer sends an acknowledgement to its external environment.
5. The robot moves to conveyer 2 with an opened clamshell while the leaving conveyer delivers the work piece to the next station.
6. The robot takes one work piece B and moves to the leaving conveyer. A new work piece A is put at the same time on conveyer 1 externally. Conveyer 1 moves until the first work piece reaches its desired position such that is can be taken by the robot.
7. The robot deposits the taken work piece B on the leaving conveyer. It waits with an opened clamshell at that position for new instructions from the control unit. The leaving conveyer sends an acknowledgement to its external environment.
8. The robot moves to conveyer 1 with an opened clamshell while the leaving conveyer delivers the work piece to the next station.
9. The procedure is continued with step 3.

Conveyer model. The models of conveyers 1 and 2 as well as the leaving conveyer relay on the following consideration:

- The power supply of each conveyer can be turned on and off by the environment.
- The motors can be started and stopped internally.

- Work pieces A and B can be put on the corresponding conveyer externally.
- The robot deposits work piece on the leaving conveyer. This action is internal.
- The robot takes work piece A and B from the corresponding conveyer. This action is internal.
- Outputting a work piece from the leaving conveyer is triggered externally.
- A new work piece can be put on conveyer 1 and 2, only if one work piece is delivered by the leaving conveyer to the next station.

The set of states is given in Appendix B.2.1 in Table B.22, the sets of control signals in Tables B.23 and the sets of interconnection signals in Table B.24 where the indices "$c1$" and "$c2$" are used for conveyers 1 and 2 as well as the index "lc" for the leaving conveyer. The behavioral relation of the conveyers 1 and 2 is shown in Table B.25(a), the behavioral relation of the leaving conveyer in Table B.25(b).

Robot model. The model of the robot bases on the following considerations:

- The power supply of the robot is turned on and off by the environment.
- It does not output anything to the environment.
- The robot can be forced to move only internally by the control unit.
- The clamshell closes to take a work piece from a conveyer and opens to put the work piece on the leaving conveyer.
- The first conveyer to which the robot moves must be defined preliminary.
- The robot reports the fulfillment of a given task to the control unit.

The set of states is given in Appendix B.2.2 in Table B.26, the sets of control signals in Tables B.27 and the sets of interconnection signals in Table B.28. The behavioral relation of the robot is shown in Table B.29.

Control unit model. The model of the control unit depends on the sorting task it has to fulfill. Work pieces A and B are delivered in this example in an interchanging order by the leaving conveyer to the next station. In addition, the following considerations hold for the control unit:

- The power supply of the control unit is turned on by the environment without any response. Turning off is not considered in this example for simplicity.

- The control unit gets information about the states of the components of the system internally. Depending on this information, it generates the next command based on the sorting program.

The set of states and control signals are given in Appendix B.2.3 in Table B.30, the sets of interconnection inputs in Table B.31 and the set of interconnection outputs in Table B.32. The behavioral relation of the control unit is shown in Table B.33.

8.2.2 Simulation

The described sorting task can be simulated with the sequence of control inputs given in Table 8.6 where v_{cu} denotes the control input of the control unit, v_{ro} of the robot, v_{lc} of the leaving conveyer, v_{c2} of conveyer 2 and v_{c1} of conveyer 1. The simulated state sequence is given in Table 8.7 and the simulated sequence of control outputs in Table 8.8.

Table 8.6: Sequence of control inputs to simulate the sorting task

k_e	0	1	2	3	4	5	6	7	8	9	10
v_{cu}	1	ε	ε	ε	ε	ε	ε	ε	ε	ε	ε
v_{ro}	1	ε	ε	ε	ε	ε	ε	ε	ε	ε	ε
v_{lc}	1	ε	ε	ε	ε	3	ε	ε	3	ε	ε
v_{c2}	1	ε	2	ε	ε	ε	ε	ε	2	ε	ε
v_{c1}	1	ε	ε	ε	ε	2	ε	ε	ε	ε	ε

Table 8.7: Simulated sequence of states

k_e	−1	0	1	2	3	4	5	6	7	8	9	10
z_{cu}	1	2	3	3	3	3	3	3	3	3	3	3
z_{ro}	1	2	5	7	3	6	8	4	5	7	3	6
z_{lc}	1	2	3	3	3	3	3	3	3	3	3	3
z_{lc}	1	2	3	3	3	3	3	3	3	3	3	3
z_{lc}	1	2	3	3	3	3	3	3	3	3	3	3

Table 8.8: Simulated sequence of control outputs

k_e	0	1	2	3	4	5	6	7	8	9	10
w_{cu}	ε	ε	ε	ε	ε	ε	ε	ε	ε	ε	ε
w_{ro}	ε	ε	ε	ε	ε	ε	ε	ε	ε	ε	ε
w_{lc}	ε	ε	ε	2	ε	3	2	ε	3	2	ε
w_{c2}	ε	ε	2	ε	ε	3	ε	ε	2	ε	ε
w_{c1}	ε	ε	3	ε	ε	2	ε	ε	3	ε	ε

Time $k_e = 0$. All components move from their initial state 1 (power supply off) to the state 2 (power supply on) due to the initialization control command $\boldsymbol{v}(0) = \mathbf{1}$ simultaneously. This synchronization arises from the isochronous change of all control inputs and not from an interaction between the components. Internally, all components behave asynchronously because the value of interconnection signals is ε for the initialization command. Hence, they can be turned on one after the other without any restrictions.

Time $k_e = 1$. The next step corresponds to the starting procedure of all components. As all components get the vanishing control input $\boldsymbol{v}(1) = \boldsymbol{\varepsilon}$, their state transitions are caused by the free motion of the control unit via the interconnection signals. Hence, all components move synchronously. The control unit and all conveyers are in their operating state $z_{cu}(1:10) = 3$, $z_{lc}(1:10) = 3$, $z_{c2}(1:10) = 3$, $z_{c1}(1:10) = 3$ from now on. The robot moves to conveyer $c1$ ($z_{ro}(1) = 5$). The control unit and the robot never influence their environment: They generate the empty control output symbol $w_{cu}(0:10) = \varepsilon$ and $w_{ro}(0:10) = \varepsilon$ at each time step.

Time $k_e = 2$. Conveyer 2 gets a new work piece B at $k_e = 2$ ($v_{c2}(2) = 2 = w_{c2}(2)$) while the robot takes a work piece A ($z_{ro}(2) = 7$). The motion of the robot is caused via the interconnection signals by the state transition of conveyer 2 which is forced externally.

Time $k_e = 3$. The robot moves to the leaving conveyer ($z_{ro}(3) = 3$) and places the work piece A. There is no interaction with conveyers 1 and 2. Hence, the robot, the control unit and the leaving conveyer perform an asynchronous state transition due to the free motion of the control unit because all control inputs are vanishing ($\boldsymbol{v}(3) = \boldsymbol{\varepsilon}$).

Time $k_e = 4$. The robot is forced by the control unit to move to conveyer $c2$ with an opened clamshell ($z_{ro}(4) = 6$). The conveyers are not involved. Consequently, the robot and the control unit perform an asynchronous state transition based on the free motion of the control unit ($\boldsymbol{v}(4) = \boldsymbol{\varepsilon}$).

Time $k_e = 5$. The next step corresponds to the transport of a work piece B by the robot ($z_{ro}(5) = 8$) to the leaving conveyer which transports the work piece A to the next work station ($v_{lc}(5) = 3 = w_{lc}$ and $w_{c2}(5) = 3$). A new work piece A is placed on conveyer $c1$ ($v_{c1}(5) = 2 = w_{c1}(5)$). Thus, all components move synchronously.

Time $k_e = 6$. The robot places the work piece B on the leaving conveyer ($z_{ro}(6) = 4$) due to a command from the control unit ($\boldsymbol{v}(6) = \boldsymbol{\varepsilon}$). The leaving conveyer communicates to its environment that it gets a new work piece by $w_{lc}(6) = 2$. In this case, the robot, the control unit and the leaving conveyer perform an asynchronous state transition, again.

Time $k_e = 7$. The robot is forced internally ($\boldsymbol{v}(7) = \boldsymbol{\varepsilon}$) to move to conveyer $c1$ with an opened clamshell ($z_{ro}(7) = 5$) by the control unit. Thus, the robot and the control unit move asynchronously with respect to the conveyers.

Time $k_e = 8$. Delivering work piece B to the next work station by the leaving conveyer is indicated by $v_{lc}(8) = 3 = w_{lc}(8)$, $v_{c1}(8) = 3$. The robot transports one work piece A to the leaving conveyer ($z_{ro}(8) = 7$). The simultaneous placement of a new work piece B in conveyer $c2$ is denoted and acknowledged by $v_{c2}(8) = 2 = w_{c2}(8)$. The first eradication round of the sorting program is finished.

Times $k_e = 9$ and $k_e = 10$. The sorting program is continued with the actions described in steps $k_e = 3$ and $k_e = 4$.

Chapter 9

Conclusion and outlook

9.1 Conclusion

This thesis has presented a novel approach to the modeling, analysis and diagnosis of coupled mechatronical systems which consist of a set of interconnected components allowing for *partial autonomy* and *asynchronous* state changes. The systems under consideration have the following properties: The internal interactions are immeasurable but reliable and the measurements relevant for diagnosis are given as a sequence of events even though the underlying control is assumed to be continuous.

Component-oriented modeling. *Asynchronous networks of I/O-automata* have been developed in this thesis to cope with partial coupling between components and to reduce the *computational complexity* of the diagnostic algorithms. I/O-automata have been used as component models. Faults have been modeled as an additional parameter of their behavioral relation. The components' measurable inputs and outputs have been modeled as control signals. Interconnection signals have been used to model the internal dependencies among the components. These interconnection signals have been linked via an interaction block to explicitly model the system structure.

The empty symbol ε has been introduced to model the non-interaction between components, if the desired task is fulfilled by the components separately. Consequently, the sequences of all local measurements are of equal length and the behavior of all components is synchronized upon the occurrence of events.

This modeling concept has been based on the idea of parallel composition known from standard automata. It has been shown that the new modeling formalism is applicable for at least the same class of asynchronous discrete-event systems as the known modeling formalism while having certain advantages.

Analysis. Coupling of component models can result in loops which may not always be *well-posed*. The criterion known from synchronous networks of I/O-automata has been extended to check for the well-posedness of the developed models.

To check for partially autonomous behavior, two types of autonomy have been introduced in this thesis. Conditions for *structural autonomy* have been derived based on a graph theoretical analysis of the corresponding directed graph. *State-dependent autonomy* has been investigated in terms of the behavioral relation and used for simplifications in diagnosis.

Two algorithms for the *simulation* of the behavior of asynchronous networks of I/O-automata with identical results have been presented. The first one has used the whole equivalent I/O-automaton which has been pre-computed by composition of the network whereas the second has reduced the model size by applying the composition rule only to the part of the behavioral relation concerning the actual state and inputs. This approach has been referred to as online composition.

Diagnosis. To carry out diagnosis, three different information structures have been investigated in this thesis: Centralized, decentralized and partially coordinated. The centralized approach has been proven to yield the ideal diagnostic result, but reduction of the computational complexity by using online composition is rather small. It has already been applied to synchronous networks of I/O-automata. The modeling introduced in this thesis makes it applicable to both, asynchronous and synchronous networks of I/O-automata as composition leads to the loss of network effects.

To reduce the computational complexity further, the diagnostic task has been solved in a decentralized way next. The locally obtained diagnostic results have been synchronized upon the occurrence of events. They have been combined to the decentrally obtained diagnostic result which has been proven to be *complete* and *sound* for well-posed networks in the case of state-dependent autonomy. A lack of soundness arises in the general case, whereas the completeness property is still given.

To overcome the soundness issue, *partially coordinated diagnosis* has been introduced: The ideal diagnostic result is obtained by removing contradictions from the locally estimated sets of interconnection signals and faults in a centralized coordinator. The result of coordination has not to be sent back to the local diagnostic units for well-posed networks in the case of state-dependent autonomy as opposed to the bidirectional algorithm known from synchronous networks of I/O-automata. The partially coordinated diagnostic result has to be stored in the coordinator for the use in the next time step. The developed diagnostic algorithm is applicable to synchronous networks of I/O-automata due to the use of the empty symbol.

9.2 Outlook

Future work related to this thesis is two-stage. The first stage is concerned with the diagnosability of an ideal system. The second state extends the diagnostic algorithm to unreliable communication.

Diagnosability. This system property specifies the success of detecting and identifying faults in the system. It has been investigated in [84] for synchronous networks of I/O-automata. Their criteria should be applicable to asynchronous networks of I/O-automata due to the use of the empty symbol. The more challenging task is to *efficiently check* the diagnosability because this check cannot be solved solely based on the component models. Reducing this computational complexity has been developed for the diagnoser approach based on efficient algorithms [54, 83, 134, 138] and verifiers [91, 129, 135]. These approaches can be used in conjunction with the criteria known from [84] to obtain a computable algorithm for the diagnosability analysis of asynchronous networks of I/O-automata.

Communication channels. A *reliable communication* between the components, the local diagnostic units and the coordinator has been assumed throughout this work in the sense that the global ordering of local measurements is preserved under communication. This assumption has to be relaxed, if data may be received with a finite or even infinite delay. Approaches to handle these uncertainties have been published for coupled standard automata using the diagnoser approach [35, 92, 96, 97] or the net unfolding approach [19, 38, 47]. In the context of the multi-model approach, communication effects have only been considered in [110, 111] for the remote diagnosis of a single I/O-automaton. Their approach can be combined with results known from coupled standard automata to account for unreliable communication in asynchronous networks of I/O-automata.

Bibliography

Contributions by the author

[1] S. Drüppel. *Vergleich der zentralen und der dezentralen Diagnose von Standardautomaten.* Internal report, Robert Bosch GmbH, 2006.

[2] S. Drüppel. *E/A-Interpretation von gekoppelten Standardautomaten.* Internal report, Robert Bosch GmbH, 2007.

[3] S. Drüppel and J. Lunze. Completeness properties of modular diagnosis of finite state machines. In *5th Workshop on Advanced Control and Diagnosis*, Grenoble, 2007.

[4] S. Drüppel, J. Lunze, and M. Fritz. Application of a method for modular diagnosis to vehicle dynamics control systems. In *4. Fachtagung für Steuerung und Regelung von Fahrzeugen und Motoren*, pages 355–364, Baden-Baden, 2008.

[5] S. Drüppel, J. Lunze, and M. Fritz. Modeling of asynchronous discrete-event systems as networks of input-output automata. In *17th IFAC World Congress*, Seoul, 2008.

[6] Y. S. Nke, S. Drüppel, and J. Lunze. Direct feedback in asynchronous networks of input-output automata. In *10th European Control Conference*, Budapest, 2009.

Supervised theses and internships

[7] S. Adelt. *Implementierung einer neuen Methodik zur Modellierung ereignisdiskreter Systeme in MATLAB/SIMULINK.* Internship report, Robert Bosch GmbH, 2008.

[8] T. Arndt and C. Maul. *Optimierung und Erweiterung der Toolbox zur Komposition und Simulation von Automatennetzen und Implementierung einer Benutzeroberfläche.* Internship report, Robert Bosch GmbH, 2008.

[9] C. Maul. *Implementierung und Erprobung eines Diagnoseverfahrens für Automatennetze.* Study thesis, Lehrstuhl für Automatisierungstechnik und Prozessinformatik, Ruhr-Universität Bochum, 2008.

[10] Y. S. Nke. *Implementierung einer neuen Methodik zur Modellierung ereignisdiskreter Systeme in MATLAB/SIMULINK*. Internship report, Robert Bosch GmbH, 2007.

[11] Y. S. Nke. *Entwicklung, Implementierung und Erprobung eines Kompositions- und Simulationsverfahrens für Automatennetze*. Diploma thesis, Lehrstuhl für Automatisierungstechnik und Prozessinformatik, Ruhr-Universität Bochum, 2008.

General literature

[12] D. Ammon. *Modellbildung und Systementwicklung in der Fahrzeugdynamik*. B. G. Teubner, Stuttgart, 1997.

[13] V. Armant, P. Dague, and L. Simon. Distributed consistency-based diagnosis. In *19th International Workshop on Principles of Diagnosis*, Blue Mountains, NSW, 2008.

[14] J. Arámburo-Lizárraga, A. Ramírez-Treviño, E. López-Mellado, and E. Ruiz-Beltrán. *Advances in Robotics, Automation and Control*, chapter Fault Diagnosis in Discrete Event Systems Using Interpreted Petri Nets, pages 71–84. InTech, 2008.

[15] E. Athanasopoulou and C. N. Hadjicostis. Decentralized failure diagnosis in discrete event systems. In *American Control Conference*, Minneapolis, Minnesota, 2006.

[16] E. Athanasopoulou, L. Li, and C. N. Hadjicostis. Maximum likelihood failure diagnosis in finte state machines under unreliable observations. *IEEE Transactions on Automatic Control*, 55(3):579–593, 2010.

[17] S. Balemi. Input/output discrete event processes and communication delays. *Discrete Event Dynamic Systems: Theory and Application*, 4:41–85, 1994.

[18] J. C. Basilio and S. Lafortune. Robust codiagnosability of discrete event systems. In *American Control Conference*, St. Louis, Missouri, 2009.

[19] A. Benveniste, E. Fabre, S. Haar, and C. Jard. Diagnosis of asynchronous discrete-event systems: A net unfolding approach. *IEEE Transactions on Automatic Control*, 48(5):714–727, 2003.

[20] M. Blanke, M. Kinnaert, J. Lunze, and M. Staroswiecki. *Diagnosis and Fault-Tolerant Control*. Springer Verlag, Berlin-Heidelberg, 2nd edition, 2006.

[21] R. K. Boel and van Schuppen J. H. Decentralized failure diagnosis for discrete-event systems with costly communication between diagnosers. In *6th International Workshop on Discrete Event Systems*, pages 175–181, Zaragoza, 2002.

[22] D. Brand and P. Zafiropulo. On communicating finite-state machines. *Journal of the Association for Computing Machinery*, 30(2):323–342, 1983.

[23] R. Bukharaev. *Theorie der stochastischen Automaten.* B. G. Teubner, 1995.

[24] M. P. Cabasino, A. Giua, and C. Seatzu. Diagnosis of discrete event systems using labeled petri nets. In *2nd IFAC Workshop on Dependable Control of Discrete Systems*, Bari, 2009.

[25] M. P. Cabasino, A. Giua, and C. Seatzu. Fault detection for discrete event systems using petri nets with unobservable transitions. *Automatica*, 46:1531–1539, 2010.

[26] L. K. Carvalho, J. C. Basilio, and M. V. Moreira. Robust diagnosability of discrete event systems subject to intermittent sensor failures. In *10th International Workshop on Discrete Event Systems*, Berlin, 2010.

[27] C. G. Cassandras and S. Lafortune. *Introduction to Discrete Event Systems.* Springer Verlag, USA, 2nd edition, 2008.

[28] O. Contant, S. Lafortune, and D. Teneketzis. Diagnosability of discrete event systems with modular structure. *Discrete Event Dynamic Systems*, 16:9–37, 2006.

[29] M.-O. Cordier and A. Grastien. Exploiting independence in a decentralised and incremental approach of diagnosis. In *17th International Workshop on Principles of Diagnosis*, Peñaranda de Duero, 2006.

[30] M. J. Daigle. *A Qualitative Event-Based Approach to Fault Diagnosis of Hybrid Systems.* PhD thesis, Graduate School of Vanderbilt University, 2008.

[31] M. J. Daigle, I. Roychoudhury, G. Biswas, and X. Koutsoukos. An event-based approach to distributed diagnosis of continuous systems. In *21st International Workshop on Principles of Diagnosis*, Portland, Oregon, 2010.

[32] J. D. de Kleer and B. Williams. Diagnosing multiple faults. *Artificial Intelligence*, 32:97–130, 1987.

[33] R. I. Debouk. *Failure Diagnosis of Decentralized Discrete Event Systems.* PhD thesis, The University of Michigan, 2000.

[34] R. I. Debouk, S. Lafortune, and D. Teneketzis. Coordinated decentralized protocols for failure diagnosis of discrete event systems. *Discrete Event Dynamic Systems: Theory and Application*, 10:33–93, 2000.

[35] R. I. Debouk, S. Lafortune, and D. Teneketzis. On the effect of communication delays in failure diagnosis of decentralized discrete event systems. In *38th IEEE Conference on Decision and Control*, Sydney, NSW, 2000.

[36] P. Dömösi and C. L. Nehaniv. *Algebraic Theory of Automata Networks.* Siam, Philadelphia, 2005.

[37] S. Even. *Graph Algorithms.* Computer Science Press, Potomac, Maryland, 1979.

[38] E. Fabre, A. Benveniste, S. Haar, and C. Jard. Distributed monitoring of concurrent and asynchronous systems. *Discrete Event Dynamic Systems: Theory and Application*, 15:33–84, 2005.

[39] D. Förstner. *Qualitative Modellierung für die Prozeßdiagnose und deren Anwendung auf Dieseleinspritzsysteme.* PhD thesis, Technische Universität Hamburg-Harburg, 2001.

[40] C. Fritsch and J. Lunze. Complexity reduction of remote diagnosis of discrete-event systems. In *17th International Symposium on Mathematical Theory of Networks and Systems*, Kyoto, 2006.

[41] C. Fritsch and J. Lunze. Remote diagnosis of discrete-event systems with on-board and off-board components. In *6th IFAC Symposium on Fault Detection, Supervision and Safety of Technical Processes*, Beijing, 2006.

[42] E. García Moreno, A. Correcher Salvador, F. Morant Anglada, E. Quiles Cucarella, and R. Blasco Giménez. Centralized modular diagnosis and the phenomenon of coupling. *Discrete Event Dynamic Systems*, 16(3):311–326, 2006.

[43] H. E. Garcia and T.-S. Yoo. Model-based detection of routing events in discrete flow networks. *Automatica*, 41:583–594, 2005.

[44] S. Genc. *On Diagnosis and Predictability of Partially-Observed Discrete-Event Systems.* PhD thesis, The University of Michigan, 2006.

[45] S. Genc and S. Lafortune. Distributed diagnosis of place-bordered petri nets. *IEEE Transactions on Automation Science and Engineering*, 4(2):206–219, 2007.

[46] R. H. Güting. *Datenstrukturen und Algorithmen.* Teubner, Stuttgart, 1992.

[47] S. Haar. Types of asynchronous diagnosability and the receals-relation in occurrence nets. *IEEE Transactions on Automatic Control*, 55(10):2310–2320, 2010.

[48] S. Hashtrudi Zad, R. H. Kwong, and W. M. Wonham. Fault diagnosis in timed discrete-event systems. In *38th Conference on Decision and Control*, Phoenix, AZ, 1999.

[49] S. Hashtrudi Zad, R. H. Kwong, and W. M. Wonham. Fault diagnosis in discrete-event systems: Framework and model reduction. *IEEE Transactions on Automatic Control*, 48(7):1199–1212, 2003.

[50] M. Heymann. Concurrency and discrete event control. *IEEE Control Systems Magazine*, 10:103–112, 1990.

[51] J. E. Hopcroft, R. Motwani, and J. D. Ullman. *Introduction to Automata Theory, Languages and Computation.* Addison-Wesley, Amsterdam, third edition, 2007.

[52] H. Hu, A. Gehin, and M. Bayart. An extended qualitative multi-faults diagnosis from first principles I: Theory and modelling. In *28th Chinese Control Conference*, Shanghai, 2009.

[53] H. Hu, A. Gehin, and M. Bayart. An extended qualitative multi-faults diagnosis from first principles II: Algorithm and case study. In *28th Chinese Control Conference*, Shanghai, 2009.

[54] S. Jiang, Z. Huang, V. Chandra, and R. Kumar. A polynomial algorithm for testing diagnosability of discrete-event systems. *IEEE Transactions on Automatic Control*, 46(8):1318–1321, 2001.

[55] S. Jiang and R. Kumar. Diagnosis of repeated failures for discrete event systems with linear-time temporal-logic specifications. *IEEE Transactions on Automation Science and Engineering*, 3(1):47–59, 2006.

[56] S. Jiang, R. Kumar, and H. E. Garcia. Optimal sensor selection for discrete-event systems with partial observation. *IEEE Transactions on Automatic Control*, 48(3):369–381, 2003.

[57] G. Jiroveanu and R. K. Boel. Petri net model-based distributed diagnosis for large interacting systems. In *16th International Workshop on Principles of Diagnosis*, Pacific Grove, CA, 2005.

[58] P. Kan John, A. Grastien, and Y. Pencolé. Synthesis of a distributed and accurate diagnoser. In *21st International Workshop on Principles of Diagnosis*, Portland, Oregon, 2010.

[59] M. Knoop, E. Liebemann, and W. Schröder. Functional architectures for vehicle dynamics management. In *VDI-Berichte*, 1931, pages 789–798, Düsseldorf, 2006.

[60] J. Kohl and A. Bauer. Role-based diagnosis for distributed vehicle functions. In *21st International Workshop on Principles of Diagnosis*, Portland, Oregon, 2010.

[61] S. Lafortune, D. Teneketzis, M. Sampath, R. Sengupta, and K. Sinnamohideen. Failure diagnosis of dynamic systems: An approach based on discrete event systems. In *American Control Conference*, Arlington, VA, 2001.

[62] G. Lamperti and M. Zanella. *Diagnosis of Active Systems.* Kluwer Academic Publishers, Dordrecht, Netherlands, 2003.

[63] E. A. Lee and P. Varaiya. *Structure and Interpretation of Signals and Systems.* Addison-Wesley, Boston, 2003.

[64] D. Lefebvre and C. Delherm. Diagnosis of DES with petri net models. *IEEE Transactions on Automation Science and Engineering*, 4(1):114–118, 2007.

[65] E. Lefebvre, D.and Leclercq. Stochastic petri net identification for the fault detection and isolation of discrete event systems. *IEEE Transactions on Systems, Man and Cybernetics, Part A: Systems and Humans*, 41(2):213–225, 2011.

[66] S. T. S. Lima, J. C. Basilio, S. Lafortune, and M. V. Moreira. Robust diagnosis of discrete-event systems subject to permanent sensor failures. In *10th International Workshop on Discrete Event Systems*, Berlin, 2010.

[67] F. Lin, K. Rudie, and S. Lafortune. Minimal communication for essential transitions in a distributed discrete-event system. *IEEE Transactions on Automatic Control*, 52(8):1495–1502, 2007.

[68] C. Lopez-Varela, A. Subias, and M. Combacau. A consistency based approach to deal with modeling errors and process failures in D.E.S. In *IEEE International Conference on Systems, Man and Cybernetics*, San Antonio, Texas, 2009.

[69] J. Lunze. Diagnosis of quantized systems based on a timed discrete-event model. *IEEE Transactions on Systems, Man and Cybernetics, Part A: Systems and Humans*, 30(3):322–335, 2000.

[70] J. Lunze. *Ereignisdiskrete Systeme*. Oldenbourg, München Wien, 2006.

[71] J. Lunze. *Automatisierungstechnik*. Oldenbourg, München Wien, 2nd edition, 2008.

[72] J. Lunze. Decentralised diagnosis of quantised systems described by asynchronous I/O automata networks. Forschungsbericht 2008.12, Lehrstuhl für Automatisierungstechnik und Prozessinformatik, Ruhr-Universität Bochum, 2008.

[73] J. Lunze. Relations between networks of standard automata and networks of I/O-automata. In *9th International Workshop on Discrete Event Systems*, Göteborg, 2008.

[74] J. Lunze. Determination of distinguishing input sequences for the diagnosis of discrete-event systems. In *2nd IFAC Workshop on Dependable Control of Discrete Systems*, Bari, 2009.

[75] J. Lunze. Diagnosability of deterministic I/O-automata. In *7th IFAC Symposium on Fault Detection, Supervision and Safety of Technical Processes*, Barcelona, 2009.

[76] J. Lunze. *Künstliche Intelligenz für Ingenieure*. Oldenbourg, München Wien, 2010.

[77] J. Lunze and J. Neidig. A new approach to distributed diagnosis using stochastic automata networks. In *1st Workshop on Advanced Control and Diagnosis*, pages 211–216, Duisburg, 2003.

[78] J. Lunze and J. Schröder. Fault diagnosis of stochastic automata networks. In *5th IFAC Symposium on Fault Detection, Supervision and Safety of Technical Processes*, Washington D.C., 2003.

[79] N. A. Lynch. I/O automata: A model for discrete event systens. In *Annual Conference on Information Sciences and Systems*, pages 29–38, Princeton, N.J., 1988.

[80] G. H. Mealy. A method for synthesizing sequential ciruits. *Bell System Technical Journal*, 34(9):1045–1079, 1955.

[81] A. Mohammadi Idghamishi and S. Hashtrudi Zad. Fault diagnosis in hierachical discrete-event systems. In *43rd IEEE Conference on Decision and Control*, Paradise Island, The Bahamas, 2004.

[82] E. F. Moore. Gedanken-experiment on sequential machines. *Auomata Studies*, 34:129–153, 1956.

[83] M. V. Moreira, T. C. Jesus, and J. C. Basilio. Polynomial time verification of decentralized diagnosability of discrete event systems. In *American Control Conference*, Baltimore, Maryland, 2010.

[84] J. Neidig. *An Automata Theoretic Approach to Modular Diagnosis of Discrete-Event Systems*. Books on Demand GmbH, Norderstedt, 2007.

[85] J. Neidig and J. Lunze. Decentralised diagnosis of automata networks. In *16th IFAC World Congress*, Prague, 2005.

[86] J. Neidig and J. Lunze. Direct feedback in automata networks. In *16th IFAC World Congress*, Prague, 2005.

[87] J. Neidig and J. Lunze. Coordinated diagnosis of nondeterministic automata networks. In *6th IFAC Symposium on Fault Detection, Supervision and Safety of Technical Processes*, Beijing, 2006.

[88] J. Neidig and J. Lunze. Unidirectional coordinated diagnosis of automata networks. In *17th International Symposium on Mathematical Theory of Networks and Systems*, Kyoto, 2006.

[89] P. Nenninger, S. Brummund, and U. Kiencke. Detection in distributed automotive electronic systems using hierachical colored bayesian petri-nets. In *SAE World Congress*, Warrendale, PA, 2005.

[90] A. Paoli and S. Lafortune. Diagnosability analysis of a class of hierarchical state machines. *Discrete Event Dynamic Systems*, 18:385–413, 2008.

[91] Y. Pencolé. Diagnosability analysis of distributed discrete event systems. In *15th International Workshop on Principles of Diagnosis*, Carcassonne, 2004.

[92] Y. Pencolé and M.-O. Cordier. A formal framework for the decentralised diagnosis of large scale discrete event systems and its application to telecommunication networks. *Artificial Intelligence*, 164(1 - 2):121–170, 2005.

[93] A. Philippot, M. Sayed-Mouchaweh, and V. Carré-Ménétrier. Discrete event model-based approach for fault detection and isolation of manufacturing systems. In *2nd IFAC Workshop on Dependable Control of Discrete Systems*, Bari, 2009.

[94] G. Provan. Distributed diagnosability properties of discrete event systems. In *American Control Conference*, Anchorage, Alaska, 2002.

[95] W. Qiu and R. Kumar. Decentralized failure diagnosis of discrete event systems. *IEEE Transactions on Systems, Man and Cybernetics, Part A: Systems and Humans*, 36(2):384–395, 2006.

[96] W. Qiu and R. Kumar. On decidability of distributed diagnosis under unbounded-delay communication. *IEEE Transactions on Automatic Control*, 52(1):114–116, 2007.

[97] W. Qiu and R. Kumar. Distributed diagnosis under bounded-delay communication of immediately forwarded local observations. *IEEE Transactions on Systems, Man and Cybernetics, Part A: Systems and Humans*, 38(3):628–643, 2008.

[98] W. Qiu, Q. Wen, and R. Kumar. Decentralized diagnosis of event-driven systems for safely reacting to failures. *IEEE Transactions on Automation Science and Engineering*, 6(2):362–366, 2009.

[99] A. Ramírez-Treviño, E. Ruiz-Beltrán, I. Rivera-Rangel, and E. López-Mellado. Online fault diagnosis of discrete event systems. a petri net-based approach. *IEEE Transactions on Automation Science and Engineering*, 4(1):31–39, 2007.

[100] R. Reiter. A theory of diagnosis from first principles. *Artificial Intelligence*, 32(1):57–95, 1987.

[101] H. Ressencourt, L. Trave-Massuyes, and J. Thomas. Hierarchical modelling and diagnosis for embedded systems. In *6th IFAC Symposium on Fault Detection, Supervision and Safety of Technical Processes*, Beijing, 2006.

[102] S. L. Ricker and J. H. van Schuppen. Decentralized failure diagnosis with asynchronous communication between supervisors. In *6th European Control Conference*, pages 1002–1006, Porto, 2001.

[103] Robert Bosch GmbH. *Kraftfahrtechnisches Taschenbuch.* Vieweg & Teubner, 27th edition, 2010.

[104] K. R. Rohloff. The diagnosis of failures via the combination of distributed observations. In *13th Mediterranean Conference on Control and Automation*, Limassol, 2005.

[105] M. Roth, J.-J. Lesage, and L. Litz. Black-box identification of discrete event systems with optimal partitioning of concurrent subsystems. In *American Control Conference*, Baltimore, Maryland, 2010.

[106] M. Sampath. *A Discrete Event Systems Approach to Failure Diagnosis.* PhD thesis, The University of Michigan, 1995.

[107] M. Sampath, R. Sengupta, S. Lafortune, K. Sinnamohideen, and D. Teneketzis. Diagnosability of discrete-event systems. *IEEE Transactions on Automatic Control*, 40:1555–1575, 1995.

[108] M. Sampath, R. Sengupta, S. Lafortune, K. Sinnamohideen, and D.C. Teneketzis. Failure diagnosis using discrete-event models. *IEEE Transactions on Control Systems Technology*, 4(2):105–124, 1996.

[109] M. Sayed-Mouchaweh. Decentralized fault detection and isolation of manufacturing systems. In *21st International Workshop on Principles of Diagnosis*, Portland, Oregon, 2010.

[110] T. Schlage and J. Lunze. Modelling of networked systems for remote diagnosis. In *Conference on Control and Fault-Tolerant Systems*, Nice, 2010.

[111] T. Schlage and J. Lunze. Remote diagnosis of timed I/O-automata. In *21st International Workshop on Principles of Diagnosis*, Portland, Oregon, 2010.

[112] G. Schmidt and T. Ströhlein. *Relationen und Graphen.* Springer, Berlin Heidelberg, 1989.

[113] J. Schröder. *Modelling, State Observation and Diagnosis of Quantised Systems.* Springer-Verlag, Berlin, 2003.

[114] G. Schullerus and V. Krebs. Diagnosis of a class of discrete event systems based on parameter estimation of a modular agebraic model. In *12th International Workshop on Principles of Diagnosis*, Sansicario, 2001.

[115] A. Schumann, Y. Pencolé, and S. Thiébaux. A spectrum of symbolic on-line diagnosis approaches. In *22nd American National Conference on Artificial Intelligence*, Vancouver, 2007.

[116] A. Schumann, Y. Pencolé, and S. Thiébaux. A decentralised symbolic diagnosis approach. In *19th European Conference on Artificial Intelligence*, Lisbon, 2010.

[117] R. S. Sreenivas and B. H. Krogh. On condition/event systens with discrete realization. *Discrete Event Dynamic Systems: Theory and Application*, 1:209–236, 1991.

[118] R. Su. *Distributed Diagnosis for Discrete-Event Systems.* PhD thesis, Graduate Department of Electrical and Computer Engineeing, 2004.

[119] R. Su and W. M. Wonham. Hierarchical distributed diagnosis under global consistency. In *7th International Workshop on Discrete Event Systems*, Reims, 2004.

[120] R. Su and W. M. Wonham. Global and local consistencies in distributed fault diagnosis for discrete-event systems. *IEEE Transactions on Automatic Control*, 50(12):1923–1935, 2005.

[121] P. Supavatanakul. *Modelling and Diagnosis of Timed Discrete-Event Systems.* Shaler-Verlag, Aachen, 2004.

[122] P. Supavatanakul and G. Schullerus. A hierarchical heterogeneous approach to diagnosis of discrete-event systems. Forschungsbericht, Lehrstuhl für Automatisierungstechnik und Prozessinformatik, Ruhr-Universität Bochum, 2004.

[123] S. Takai. Robust failure diagnosis of partially observed discrete event systems. In *10th International Workshop on Discrete Event Systems*, Berlin, 2010.

[124] D. Thorsley and D. Teneketzis. Diagnosis of cyclic discrete-event systems using active acquisition of information. In *8th International Workshop on Discrete Event Systems*, Ann Arbor, Michigan, 2006.

[125] A. Trächtler. Integrated vehicle dynamics control using active brake, steering and suspension systems. *International Journal of Vehicle Design*, 36(1):1–12, 2004.

[126] A. Trächtler. Integrierte Fahrdynamikregelung mit ESP, aktiver Lenkung und aktivem Fahrwerk. *at - Automatisierungstechnik*, 53:11–19, 2005.

[127] A. Verhagen, S. Futterer, J. Rupprecht, and A. Trächtler. Vehicle dynamics management - benefits of integrated control of active brake, active steering and active suspension systems. In *FISITA World Automotive Congress*, Barcelona, 2004.

[128] W. Wang, S. Lafortune, and F. Lin. On the minimization of communication in networked systems with a central station. *Discrete Event Dynamic Systems*, 18:415–443, 2008.

[129] Y. Wang, T.-S. Yoo, and S. Lafortune. Diagnosis of discrete event systems using decentralized architectures. *Discrete Event Dynamic Systems: Theory and Application*, 17(2):233–263, 2007.

[130] Y. L. Wen, P. S. Fan, and M. D Jeng. Diagnosable discrete event systems design. In *IEEE International Conference on Systems, Man and Cybernetics*, Montréal, 2007.

[131] F. Wenck. *Modellbildung, Analyse und Steuerungsentwurf für gekoppelte ereignisdiskrete Systeme.* Shaker Verlag, Aachen, 2006.

[132] F. Wenck and J. H. Richter. A composition oriented perspective on controllability of large scale DES. In *7th International Workshop on Discrete Event Systems*, pages 271–276, Reims, 2004.

[133] W. M. Wonham. *Supervisory Control of Discrete-Event Systems.* Lecture notes, Dept. of ECE, Systems Control Group, University of Toronto, 2005.

[134] Y. Yan, L. Ye, and P. Dague. Diagnosability for patterns in distributed discrete event systems. In *21st International Workshop on Principles of Diagnosis*, Portland, Oregon, 2010.

[135] T.-S. Yoo and S. Lafortune. Polynomial-time verification of diagnosability of partially-observed discrete-event systems. *IEEE Transactions on Automatic Control*, 47:1491–1495, 2002.

[136] X. Yunxia. *Integrated Fault Diagnosis Scheme Using Finte-State Automaton.* PhD thesis, Department of Electrical and Computer Engineering, National University of Singapore, 2003.

[137] C. Zhou and R. Kumar. Computation of Diagnosable Fault-Occurrence Indices for Systems With Repeatable Faults. *IEEE Transactions on Automatic Control*, 54(7):1477–1489, 2009.

[138] C. Zhou, R. Kumar, and R. S. Sreenivas. Decentralized modular diagnosis of concurrent discrete event systems. In *9th International Workshop on Discrete Event Systems*, Göteborg, 2008.

Appendix A

Nomenclature

A.1 General notation

Type	Convention	Example
scalars	lower-case, italic	v, w
vectors	lower-case, italic, bold	$\boldsymbol{v}$, $\boldsymbol{w}$, $\boldsymbol{k}$
matrices	upper-case, italic, bold	$\boldsymbol{K}$, $\boldsymbol{G}$
sequences of scalars	upper-case, italic	V, W
sequences of sets	upper-case, italic, bold	$\boldsymbol{V}$, $\boldsymbol{W}$
relations, behavioral functions	upper case, italic, bold	$\boldsymbol{L}$
sets of scalars	upper-case, calligraphic, index regular	$\mathcal{N}_{z_i}$, $\mathcal{F}$
sets of vectors	upper-case, calligraphic, index bold	$\mathcal{N}_{\boldsymbol{z}}$, $\mathcal{N}_{\boldsymbol{v}}$

A.2 Symbols

Symbol	Description	Page
$\boldsymbol{A}$	associated matrix of a directed graph	65
$\mathcal{A}$	set of arcs (edges) of a directed graph	64
$\boldsymbol{B}$	sequence of measurements	20, 28
$\mathcal{B}$	behavior	20, 28
D	diagnostic unit	6
$\mathcal{D}$	deterministic I/O-automaton	29
$\mathcal{DAN}$	deterministic asynchronous network of I/O-automata	42
f	fault signal	20, 27
$\boldsymbol{F}$	interconnection relation	34
$\mathcal{F}$	set of possible faults (diagnostic result)	83, 95, 110,
$\mathcal{F}^\star$	set of fault candidates (ideal diagnostic result)	21
$\boldsymbol{G}$	state transition relation	34
$\mathcal{G}$	interconnection graph	64

$\boldsymbol{H}$	output relation	34
$\mathcal{J}$	coordinated set of pairs	110
k	discrete event counter	29
$\boldsymbol{k}_i^T$	i^{th} row vector of the coupling model	39, 41
$\boldsymbol{K}$	coupling model	39
$\boldsymbol{L}$	behavioral relation	30, 32, 33, 39
M	number of symbols of a control input signal	27, 30, 38
N	number of states of an I/O-automaton	30, 42
$\mathcal{N}$	nondeterministic I/O-automaton	32, 33, 39
$\mathcal{N}_\bullet$	set of symbols	20, 27, 30, 38, 42
$\mathcal{NAN}$	nondeterministic asynchronous network of I/O-automata	42
$\mathbb{N}$	set of natural numbers	20
P	number of symbols of an interconnection input signal	28, 30, 38
$\boldsymbol{poss}$	possibility distribution function	72
$Poss$	possibility function	36
r	interconnection output	28, 30, 38
R	number of symbols of a control output signal	28, 30, 38
$\mathcal{R}$	set of possible pairs sent to the coordinator	108
$\mathcal{R}'$	set of impossible pairs sent to the local diagnostic units	109
s	interconnection input	28, 30, 38
S	number of fault symbols of a fault signal	20, 27
$\mathcal{S}$	deterministic standard automaton	49
T	number of symbols of an interconnection output signal	28, 30, 38
v	control input	27, 30, 38
$\mathcal{V}$	set of vertices (nodes) of a directed graph	64
w	control output	28, 30, 38
z	state	30, 42
$\mathcal{Z}$	set of possible states (observation result)	81
$\mathcal{Z}^\star$	set of state candidates (ideal observation result)	79
$*$	asterisk, denotes any value of a given set	118
α	number of not connected, disjoint sets of strongly connected nodes	65
δ	state transition function of a standard automaton	49
ε	empty symbol	45
μ	number of control inputs of a component	38
ν	number of components of a system	27
π	number of interconnection inputs of a component	38
ρ	number of control outputs of a component	38
σ	event	25
Σ	set of events of a standard automaton	49
τ	number of interconnection outputs of a component	38
χ	characteristic function of a behavioral relation	30, 32, 33

A.3 Indices and accents

Symbol	Type	Description
d, n	superscripts	d = deterministic, n = nondeterministic, used for I/O-automata and relations
c, dc, pc	superscripts	c = centralized, dc = decentralized, pc = partially coordinated, used for the diagnostic approaches
$spur$	superscripts	$spur$ = spurious, used for faults that are no fault candidates
act	superscript	act = activated or activatable, used for the set of activated inputs or activatable outputs
$auto$	superscript	$auto$ = autonomous, used for the set of autonomous states
com, $priv$	superscript	com = common, $priv$ = privat, used to differentiate the event set of standard automata
pac	superscript	pac = parallel composition, used for the interconnection function
$\bigstar$	superscript	Ideal (diagnostic) result
T	superscript	transpose of a vector or a matrix
j	superscript	belonging to the jth signal of a component
v, w, s, r, f, z	index	v = control input, w = control output, s = interconnection input, r = interconnection output, f = fault and z = state, used for sets
i, j	index	belonging to the ith resp. j^{th} component
$\tilde{i}$, $\tilde{j}$	index	belonging to the $\tilde{i}^{th}$ resp. $\tilde{j}^{th}$ subsystem
0	index	initial value or set of initial values
e	index	number of the last element in a sequence
$\|\|$	index	result of the parallel composition
sim	index	sim = simulation, used for the simulation matrix
$\tilde{}$ (tilde)	accent	without coupling signals, used for I/O-automata, behaviors and relations
$\check{}$ (check)	accent	characteristic function, used in the case of eliminated coupling signals
$\widehat{}$ (hat)	accent	autonomous, i.e. without any input and output signals, used for I/O-automata and relations
$\bar{}$ (bar)	accent	fixed-points solving the feedback problem, characteristic function in the case of autonomy and reachability matrix of a directed graph

A.4 Operators

Operator	Meaning
$\neg$	boolean negation
$\wedge$	boolean AND (conjunction)
$\bigwedge_{i=1}^{\nu}$	evaluate the expression for $i = 1, \ldots, \nu$ and connect all with AND-operation
$\vee$	boolean OR (disjunction)
$\bigvee_{\boldsymbol{r}=1}^{T}$	evaluate the expression for $\boldsymbol{r} = 1, \ldots, T$ and connect all with OR-operation
$\times$	cartesian product of two sets
$\otimes$	Hadamard product of two vectors (element by element multiplication)
$\mathcal{N}_{\bullet}^{*}$	Kleene closure of a set of symbols
$O(\bullet)$	Landau symbol to estimate the memory consumption

Appendix B

Models of the application examples

B.1 Process plant

This section shows the discrete-event models of the components of the process plant considered in Section 8.1.

B.1.1 Vessel model

The sets of states and faults of the vessel are given in Table B.1, the control signals in Table B.2, the sets of interconnection inputs in Table B.3 and the set of interconnection outputs in Table B.4. The behavioral relation of the vessel subject to fault $f_v = 0$ is shown in Table B.5, subject to fault $f_v = 1$ in Table B.6, subject to fault $f_v = 2$ in Table B.7 and subject to fault $f_v = 3$ in Table B.8.

Table B.1: Sets of states and faults of the vessel

(a) Set of states

z_v	Meaning
1	Vessel is empty & cold
2	Vessel is empty & hot
3	Vessel is full & cold
4	Vessel is full & hot

(b) Set of faults

f_v	Meaning
0	Fault-free
1	Leakage at the bottom
2	Level sensor broken
3	Temperature sensor broken

Table B.2: Set of control signals of the vessel

(a) Set of control inputs 1

v_v^1	Meaning
ε	Ignore the input
1	Vessel is empty
2	Vessel is full

(b) Set of control inputs 2

v_v^2	Meaning
ε	Ignore the input
1	Vessel is cold
2	Vessel is hot

(c) Set of control outputs

w	Meaning
ε	Do nothing
1	Vessel is empty & cold
2	Vessel is empty & hot
3	Vessel is full & cold
4	Vessel is full & hot

Table B.3: Set of interconnection inputs of the vessel

(a) Set of interconnection inputs 1 and 2

$s_v^1 = s_v^2$	Meaning
ε	Ignore the input
1	Valve is closed
2	Valve is opened

(b) Set of interconnection inputs 3

s_v^3	Meaning
ε	Ignore the input
1	Heater is turned off
2	Heater is turned on

Table B.4: Set of interconnection outputs of the vessel

(a) Set of interconnection outputs 1 and 2

$r_v^1 = r_v^2$	Meaning
ε	Do nothing
1	Request valve position

(b) Set of interconnection outputs 3

$s_1 = s_2$	Meaning
ε	Do nothing
1	Request heater status

Table B.5: Behavioral relation of the vessel for fault $f_v = 0$

z'_v	w_v	r_v^1	r_v^2	r_v^3	z_v	v_v^1	v_v^2	s_v^1	s_v^2	s_v^3
2	2	ε	ε	1	1	ε	2	ε	ε	2
3	3	1	1	ε	1	2	ε	2	1	ε
4	4	1	1	1	1	2	2	2	1	2
1	ε	ε	ε	ε	1	ε	ε	ε	ε	ε
1	1	ε	ε	1	2	ε	1	ε	ε	1
3	3	1	1	1	2	2	1	2	1	1
4	4	1	1	ε	2	2	ε	2	1	ε
2	ε	ε	ε	ε	2	ε	ε	ε	ε	ε
1	1	1	1	ε	3	1	ε	1	2	ε
2	2	1	1	1	3	1	2	1	2	2
3	ε	ε	ε	ε	3	ε	ε	ε	ε	ε
4	4	ε	ε	1	3	ε	2	ε	ε	2
1	1	1	1	1	4	1	1	1	2	1
2	2	1	1	ε	4	1	ε	1	2	ε
3	3	ε	ε	1	4	ε	1	ε	ε	1
4	ε	ε	ε	ε	4	ε	ε	ε	ε	ε

Table B.6: Behavioral relation of the vessel for fault $f_v = 1$

z'_v	w_v	r_v^1	r_v^2	r_v^3	z_v	v_v^1	v_v^2	s_v^1	s_v^2	s_v^3
2	2	ε	ε	1	1	ε	2	ε	ε	2
1	ε	ε	ε	ε	1	ε	ε	ε	ε	ε
1	1	ε	ε	1	2	ε	1	ε	ε	1
2	ε	ε	ε	ε	2	ε	ε	ε	ε	ε
1	1	1	1	ε	3	1	ε	1	1	ε
1	1	1	1	ε	3	1	ε	2	2	ε
1	1	1	1	ε	3	1	ε	2	1	ε
1	1	1	1	ε	3	1	ε	1	2	ε
2	2	1	1	1	3	1	2	1	1	2
2	2	1	1	1	3	1	2	2	2	2
2	2	1	1	1	3	1	2	2	1	2
2	2	1	1	1	3	1	2	1	2	2
4	4	ε	ε	1	3	ε	2	ε	ε	2
3	ε	ε	ε	ε	3	ε	ε	ε	ε	ε
1	1	1	1	1	4	1	1	1	1	1
1	1	1	1	1	4	1	1	2	2	1
1	1	1	1	1	4	1	1	2	1	1
1	1	1	1	1	4	1	1	1	2	1
2	2	1	1	ε	4	1	ε	1	1	ε
2	2	1	1	ε	4	1	ε	2	2	ε
2	2	1	1	ε	4	1	ε	2	1	ε
2	2	1	1	ε	4	1	ε	1	2	ε
3	3	ε	ε	1	4	ε	1	ε	ε	1
4	ε	ε	ε	ε	4	ε	ε	ε	ε	ε

Table B.7: Behavioral relation of the vessel for fault $f_v = 2$

z'_v	w_v	r_v^1	r_v^2	r_v^3	z_v	v_v^1	v_v^2	s_v^1	s_v^2	s_v^3
2	2	ε	ε	1	1	ε	2	ε	ε	2
3	3	1	1	ε	1	2	ε	1	1	ε
3	3	1	1	ε	1	2	ε	2	2	ε
3	3	1	1	ε	1	2	ε	1	2	ε
4	4	1	1	1	1	2	2	1	1	2
4	4	1	1	1	1	2	2	2	2	2
4	4	1	1	1	1	2	2	1	2	2
1	ε	ε	ε	ε	1	ε	ε	ε	ε	ε
1	1	ε	ε	1	2	ε	1	ε	ε	1
3	3	1	1	1	2	2	1	1	1	1
3	3	1	1	1	2	2	1	2	2	1
3	3	1	1	1	2	2	1	1	2	1
4	4	1	1	ε	2	2	ε	1	1	ε
4	4	1	1	ε	2	2	ε	2	2	ε
4	4	1	1	ε	2	2	ε	1	2	ε
2	ε	ε	ε	ε	2	ε	ε	ε	ε	ε
1	1	1	1	ε	3	1	ε	1	1	ε
1	1	1	1	ε	3	1	ε	2	2	ε
1	1	1	1	ε	3	1	ε	2	1	ε
2	2	1	1	1	3	1	2	1	1	2
2	2	1	1	1	3	1	2	2	2	2
2	2	1	1	1	3	1	2	2	1	2
4	4	ε	ε	1	3	ε	2	ε	ε	2
3	ε	ε	ε	ε	3	ε	ε	ε	ε	ε
1	1	1	1	1	4	1	1	1	1	1
1	1	1	1	1	4	1	1	2	2	1
1	1	1	1	1	4	1	1	2	1	1
2	2	1	1	ε	4	1	ε	1	1	ε
2	2	1	1	ε	4	1	ε	2	2	ε
2	2	1	1	ε	4	1	ε	2	1	ε
3	3	ε	ε	1	4	ε	1	ε	ε	1
4	ε	ε	ε	ε	4	ε	ε	ε	ε	ε

Table B.8: Behavioral relation of the vessel for fault $f_v = 3$

z_v'	w_v	r_v^1	r_v^2	r_v^3	z_v	v_v^1	v_v^2	s_v^1	s_v^2	s_v^3
2	2	ε	ε	1	1	ε	2	ε	ε	1
3	3	1	1	ε	1	2	ε	2	1	ε
4	4	1	1	1	1	2	2	2	1	1
1	ε	ε	ε	ε	1	ε	ε	ε	ε	ε
1	1	ε	ε	1	2	ε	1	ε	ε	2
3	3	1	1	1	2	2	1	2	1	2
4	4	1	1	ε	2	2	ε	2	1	ε
2	ε	ε	ε	ε	2	ε	ε	ε	ε	ε
1	1	1	1	ε	3	1	ε	1	2	ε
2	2	1	1	1	3	1	2	1	2	1
4	4	ε	ε	1	3	ε	2	ε	ε	1
3	ε	ε	ε	ε	3	ε	ε	ε	ε	ε
1	1	1	1	1	4	1	1	1	2	2
2	2	1	1	ε	4	1	ε	2	1	ε
3	3	ε	ε	1	4	ε	1	ε	ε	2
4	ε	ε	ε	ε	4	ε	ε	ε	ε	ε

B.1.2 Valve model

The sets of states and faults of the valves are given in Table B.9, the control signals in Table B.10, the sets of interconnection signals in Table B.11 and the set of interconnection outputs in Table B.32. The behavioral relation of the fault-free valves are given in Table B.12 and subject to the considered faults in Table B.13.

Table B.9: Sets of states and faults of the valves

(a) Set of states

$z_{v1} = z_{v2}$	Meaning
1	Valve is closed
2	Valve is opened

(b) Set of faults

$f_{v1} = f_{v2}$	Meaning
0	Fault-free
1	Compressed air broken

Table B.10: Set of control signals of the valves

(a) Set of control inputs

$v_{v1} = v_{v2}$	Meaning
ε	Ignore the input
1	Close valve
2	Open valve

(b) Set of control outputs

$w_{v1} = w_{v2}$	Meaning
ε	Do nothing
1	Valve is closed
2	Valve is opened

Table B.11: Set of interconnection signals of the valves

(a) Set of interconnection inputs

$s_{v1} = s_{v2}$	Meaning
ε	Ignore the input
1	Vessel requests position

(b) Set of interconnection outputs

$r_{v1} = r_{v2}$	Meaning
ε	Do nothing
1	Valve is closed
2	Valve is opened

Table B.12: Behavioral relation of the fault-free valves ($f_{v1} = f_{v2} = 0$)

z'_{vi}	w_{vi}	r_{vi}	z_{vi}	v_{vi}	s_{vi}
1	ε	ε	1	ε	ε
1	1	ε	1	1	ε
1	ε	1	1	ε	1
1	1	1	1	1	1
2	2	ε	1	2	ε
2	2	2	1	2	1
1	1	ε	2	1	ε
1	1	1	2	1	1
2	2	ε	2	2	ε
2	ε	ε	2	ε	ε
2	ε	2	2	ε	1
2	2	2	2	2	1

Table B.13: Behavioral relation of the valves subject to faults

(a) Fault $f_{v1} = f_{v2} = 1$

z'_{vi}	w_{vi}	r_{vi}	z_{vi}	v_{vi}	s_{vi}
1	ε	ε	1	ε	ε
1	1	ε	1	1	ε
1	1	ε	1	2	ε
1	ε	1	1	ε	1
1	1	1	1	1	1
1	1	1	1	2	1
2	ε	ε	2	ε	ε
2	2	ε	2	1	ε
2	2	ε	2	2	ε
2	ε	2	2	ε	1
2	2	2	2	1	1
2	2	2	2	2	1

(b) Fault $f_{v1} = f_{v2} = 2$

z'_{vi}	w_{vi}	r_{vi}	z_{vi}	v_{vi}	s_{vi}
1	ε	ε	1	ε	ε
1	1	ε	1	1	ε
1	1	ε	1	2	ε
1	ε	1	1	ε	1
1	1	1	1	1	1
1	1	1	1	2	1
1	ε	ε	2	ε	ε
1	1	ε	2	1	ε
1	1	ε	2	2	ε
1	ε	1	2	ε	1
1	1	1	2	1	1
1	1	1	2	2	1

B.1.3 Heater model

The sets of states and faults of the heater are given in Table B.14, the control signals in Table B.15, the sets of interconnection signals in Table B.16 and the set of interconnection outputs in Table B.32. The behavioral relation of the heater is given in Table B.17.

Table B.14: Sets of states and faults of the heater

(a) Set of states

z_h	Meaning
1	Heater turned off
2	Heater turned on

(b) Set of faults

f_h	Meaning
0	Fault-free

Table B.15: Set of control signals of the heater

(a) Set of control inputs

v_h	Meaning
ε	Ignore the input
1	Turn heater off
2	Turn heater on

(b) Set of control outputs

w_h	Meaning
ε	Do nothing
1	Heater is turned off
2	Heater is turned on

Table B.16: Set of interconnection signals of the heater

(a) Set of interconnection inputs 1

s_h^1	Meaning
ε	Ignore the input
1	Vessel requests status

(b) Set of interconnection inputs 2

s_h^2	Meaning
ε	Ignore the input
1	Stirrer is turned off
2	Stirrer is turned on
3	Turn heater off

(c) Set of interconnection outputs 1

r_h	Meaning
ε	Do nothing
1	Stirrer is turned off
2	Stirrer is turned on

(d) Set of interconnection outputs 2

r_h	Meaning
ε	Do nothing
1	Stirrer requests status
2	Heater is turned off

Table B.17: Behavioral relation of the heater

z'_h	w_h	r^1_h	r^2_h	z_h	v_h	s^1_h	s^2_h
1	ε	ε	ε	1	ε	ε	ε
1	1	1	ε	1	1	1	ε
1	1	1	ε	1	1	2	ε
1	1	2	ε	1	1	3	ε
1	1	1	ε	1	2	1	ε
1	1	2	ε	1	2	3	ε
1	ε	ε	1	1	ε	ε	1
1	1	1	1	1	1	1	1
1	1	1	1	1	1	2	1
1	1	2	1	1	1	3	1
1	1	1	1	1	2	1	1
1	1	2	1	1	2	3	1
2	2	1	ε	1	2	2	ε
2	2	1	2	1	2	2	1
1	1	1	ε	2	1	2	ε
1	1	2	ε	2	1	3	ε
1	1	2	ε	2	2	3	ε
1	1	1	1	2	1	2	1
1	1	2	1	2	1	3	1
1	1	2	1	2	2	3	1
2	ε	ε	ε	2	ε	ε	ε
2	2	1	ε	2	2	2	ε
2	ε	ε	2	2	ε	ε	1
2	2	1	2	2	2	2	1

B.1.4 Stirrer model

The sets of states and faults of the stirrer are given in Table B.18, the control signals in Table B.19, the sets of interconnection signals in Table B.20 and the set of interconnection outputs in Table B.32. The behavioral relation of the stirrer is given in Table B.21.

Table B.18: Sets of states and faults of the stirrer

(a) Set of states

z_s	Meaning
1	Stirrer turned on
2	Stirrer turned off

(b) Set of faults

f_s	Meaning
0	Fault-free

Table B.19: Set of control signals of the stirrer

(a) Set of control inputs

v_s	Meaning
ε	Ignore the input
1	Turn stirrer off
2	Turn stirrer on

(b) Set of control outputs

w_s	Meaning
ε	Do nothing
1	Stirrer is turned off
2	Stirrer is turned on

Table B.20: Set of interconnection signals of the stirrer

(a) Set of interconnection inputs

s_s	Meaning
ε	Ignore the input
1	Heater requests status
2	Heater has turned off

(b) Set of interconnection outputs

r_s	Meaning
ε	Do nothing
1	Stirrer is turned off
2	Stirrer is turned on
3	Turn heater off

Table B.21: Behavioral relation of the stirrer

z'_s	w_s	r_s	z_s	v_s	s_s
1	ε	ε	1	ε	ε
1	1	ε	1	1	ε
1	ε	1	1	ε	1
1	1	1	1	1	1
2	2	ε	1	2	ε
2	2	2	1	2	1
1	1	3	2	1	2
2	ε	ε	2	ε	ε
2	2	ε	2	2	ε
2	2	2	2	2	1

B.2 Production facility

The discrete-event models of the components of the production facility considered in Section 8.2 are shown in this section in terms of nondeterministic I/O-automata.

B.2.1 Conveyer model

The set of states of the conveyers is given in Table B.22, the sets of control signals in Tables B.23 and the sets of interconnection signals in Table B.24 where the indices "$c1$"

and "$c2$" are used for conveyers 1 and 2 as well as the index "lc" for the leaving conveyer. The behavioral relation of the conveyers 1 and 2 is shown in Table B.25(a), the behavioral relation of the leaving conveyer in Table B.25(b).

Table B.22: Set of states of the conveyers

z_{ci}/z_{cl}	Meaning
1	Power supply off
2	Power supply on
3	Conveyer is started
4	Conveyer is stopped

Table B.23: Set of control signals of the conveyers

(a) Set of control inputs

v_{ci}/v_{lc}	Meaning
1	Turn power supply on
2	Take new work piece ($c1$ and $c2$)
3	Deliver work piece (lc)
ε	Ignore the input

(b) Set of control outputs

w_{ci}/w_{lc}	Meaning
2	New work piece taken
3	Work piece delivered
ε	Do nothing

Table B.24: Set of interconnection signals of the conveyers

(a) Set of interconnection inputs

s_{ci}/s_{lc}	Meaning
1	Start conveyer
2	Stop conveyer
3	Deliver work piece ($c1$ and $c2$)
5	Take new work piece (lc)
ε	Ignore the input

(b) Set of interconnection outputs

r_{ci}/r_{lc}	Meaning
1	Conveyer is started
2	New work piece taken
3	Work piece delivered
4	Conveyer is stopped
ε	Do nothing

Table B.25: Behavioral relation of the conveyers

(a) Conveyers 1 and 2

z'_{ci}	w_{ci}	r_{ci}	z_{ci}	v_{ci}	s_{ci}
2	ε	ε	1	1	ε
3	ε	1	2	ε	1
3	2	2	3	2	ε
3	3	3	3	ε	3
3	ε	ε	3	ε	ε

(b) Leaving conveyer

z'_A	w_A	r_A	z_A	v_A	s_A
2	ε	ε	1	1	ε
3	ε	1	2	ε	1
3	2	2	3	ε	5
3	ε	ε	3	ε	ε
3	3	3	3	3	ε

B.2.2 Robot model

The set of states of the robot is given in Table B.26, the sets of control signals in Tables B.27 and the sets of interconnection signals in Table B.28. The behavioral relation of the robot is shown in Table B.29.

Table B.26: Set of states of the robot

z_{ro}	Meaning		
	Position	Clamshell	Latest transport
1	Idle state	–	–
2	lc	opened	none
3	lc	opened	A
4	lc	opened	B
5	$c1$	opened	–
6	$c2$	opened	–
7	lc	closed with A	–
8	lc	closed with B	–

Table B.27: Set of control signals of the robot

(a) Set of control inputs

v_{ro}	Meaning
1	Turn power supply on
2	Turn power supply off
ε	Ignore the input

(b) Set of control outputs

w_{ro}	Meaning
ε	Do nothing

Table B.28: Set of interconnection signals of the robot

(a) Set of interconnection inputs

s_{ro}	Meaning
1	Move to $c1$ - open clamshell
2	Move to $c2$ - open clamshell
3	Take A - move to lc
4	Take B - move to lc
5	Deliver A - open clamshell
6	Deliver B - open clamshell
ε	Ignore the input

(b) Set of interconnection outputs

r_{ro}	Meaning
1	Moves to $c1$ - clamshell opened
2	Moves to $c2$ - clamshell opened
3	Moves to lc with A
4	Moves to lc with B
5	A delivered on lc
6	V delivered on lc
ε	Do nothing

Table B.29: Behavioral relation of the robot

(a) Part 1

z'_R	w_R	r_R	z_R	v_R	s_R
2	ε	ε	1	1	ε
5	ε	1	2	ε	1
7	ε	3	5	ε	3
8	ε	4	6	ε	4

(b) Part 2

z'_R	w_R	r_R	z_R	v_R	s_R
3	ε	5	7	ε	5
4	ε	6	8	ε	6
6	ε	2	3	ε	2
5	ε	1	4	ε	1

B.2.3 Control unit model

The set of states and control signals of the control unit are given in Table B.30, the sets of interconnection inputs in Table B.31 and the set of interconnection outputs in Table B.32. The behavioral relation of the control unit is shown in Table B.33.

Table B.30: Set of states and control signals of the control unit

(a) Set of states

z_{cu}	Meaning
1	Power supply off
2	Power supply on
3	Sorting program runs

(b) Set of control inputs

v_{cu}	Meaning
1	Turn power supply on
ε	Ignore the input

(c) Set of control outputs

w_{cu}	Meaning
ε	Do nothing

Table B.31: Set of interconnection signals of the control unit

s^1_{cu}	s^2_{cu}	s^3_{cu}	s^4_{cu}	Meaning
ε	ε	ε	ε	Ignore the input
ε	ε	ε	1	Robot moves to $c1$ with clamshell opened
ε	ε	ε	2	Robot moves to $c2$ with clamshell opened
ε	ε	2	5	lc gets A
ε	ε	2	6	lc gets B
1	1	1	1	$c1$, $c2$ and lc run, robot moves to $c1$ with clamshell opened
1	1	1	2	$c1$, $c2$ and lc run, robot moves to $c2$ with clamshell opened
3	2	3	3	$c2$ gets new B, $c1$ delivers A, robot moves with A to lc, lc delivers work piece
2	3	3	4	$c1$ gets new A, $c2$ delivers B, robot moves with B to lc, lc delivers work piece

Table B.32: Set of interconnection signals of the control unit

r^1_{cu}	r^2_{cu}	r^3_{cu}	r^4_{cu}	Meaning
ε	ε	ε	ε	Do nothing
ε	ε	ε	1	Robot shall move to $c1$ with open clamshell
ε	ε	ε	2	Robot shall move to $c2$ with open clamshell
ε	ε	5	5	Robot shall deliver A and open clamshell
ε	ε	5	6	Robot shall deliver B and open clamshell
3	ε	ε	3	Robot shall take A from $c1$ and move to lc while lc shall deliver a work piece
ε	3	ε	4	Robot shall take B from $c2$ and move to lc while lc shall deliver a work piece
1	1	1	1	$c1$, $c2$ and lc shall start, robot shall move to $c1$ with open clamshell
1	1	1	2	$c1$, $c2$ and lc shall start, robot shall move to $c2$ with open clamshell

Table B.33: Behavioral relation of the control unit

z'_{cu}	w_{cu}	r^1_{cu}	r^2_{cu}	r^3_{cu}	r^4_{cu}	z_{cu}	v_{cu}	s^1_{cu}	s^2_{cu}	s^3_{cu}	s^4_{cu}
2	ε	ε	ε	ε	ε	1	1	ε	ε	ε	ε
3	ε	1	1	1	1	2	ε	1	1	1	1
3	ε	1	1	1	2	2	ε	1	1	1	2
3	ε	3	ε	ε	3	3	ε	3	2	ε	3
3	ε	ε	3	ε	4	3	ε	2	3	ε	4
3	ε	3	ε	ε	3	3	ε	3	2	3	3
3	ε	ε	3	ε	4	3	ε	2	3	3	4
3	ε	ε	ε	5	5	3	ε	ε	ε	2	5
3	ε	ε	ε	5	6	3	ε	ε	ε	2	6
3	ε	ε	ε	ε	2	3	ε	ε	ε	ε	2
3	ε	ε	ε	ε	1	3	ε	ε	ε	ε	1